The Classical and Quantum
6j-symbols

The Classical and Quantum
6j-symbols

by

J. Scott Carter
Daniel E. Flath
and
Masahico Saito

Mathematical Notes 43

PRINCETON UNIVERSITY PRESS
———
PRINCETON, NEW JERSEY
1995

The Princeton Mathematical Notes are edited by
Luis A. Caffarelli, John N. Mather, and Elias M. Stein

Princeton University Press books are printed on acid-free paper
and meet the guidelines for permanence and durability of the
Committee on Production Guidelines for Book Longevity
of the Council on Library Resources

Library of Congress Cataloging-in-Publication Data

Carter, J. Scott.
The classical and quantum 6j-symbols / by J. Scott Carter,
Daniel E. Flath and Masahico Saito.
p. cm. — (Mathematical notes ; 43)
Includes bibliographical references.
ISBN 0-691-02730-7 (pbk. : alk. paper)
1. Quantum groups. 2. Representations of groups. I. Flath, Daniel E.
II. Saito, Masahico, 1959– . III. Title. IV. Series: Mathematical Notes
(Princeton University Press) ; 43.
QC174.17.G7C37 1995
530.1'2'0151255—dc20 95-42889

The publisher would like to acknowledge the authors of this volume
for providing the camera-ready copy from
which this book was printed

Printed in the United States of America
by Princeton Academic Press

3 5 7 9 10 8 6 4 2

Dedicated to:
Our Wives

Contents

Foreword   ix

1 Introduction   3

2 **Representations of** $U(sl(2))$ 7

    Basic definitions   7

    Finite dimensional irreducible representations   7

    Diagrammatics of $U(sl(2))$ invariant maps   12

    The Temperley-Lieb algebra   15

    Tensor products of irreducible representations   21

    The $6j$-symbols   27

    Computations   43

    A recursion formula for the $6j$-symbols   63

    Remarks   65

3 **Quantum** $sl(2)$ 67

    Some finite dimensional representations   67

    Representations of the braid groups   70

    A finite dimensional quotient of $\mathbf{C}[B(n)]$.   74

    A model for the representations V_A^j   77

    The Jones-Wentzl projectors   80

    The quantum Clebsch-Gordan theory   93

    Quantum network evaluation   99

    The quantum $6j$-symbols — generic case   106

    Diagrammatics of weight vectors (quantum case)   110

    Twisting rules   111

    Symmetries   123

    Further identities among the quantum $6j$-symbols . . .   125

4 The Quantum Trace and Color Representations 127

The quantum trace 127

A bilinear form on tangle diagrams 130

Color representations 133

The quantum $6j$-symbol — root of unity case 139

5 The Turaev-Viro Invariant 151

The definition of the Turaev-Viro invariant 151

Epilogue 157

References 160

Foreword

This book discusses the representation theory of classical and quantum $U(sl(2))$ with an eye towards topological applications of the latter. We use the Temperley-Lieb algebra and the quantum spin-networks to organize the computations. We define the $6j$-symbols in the classical, quantum, and quantum-root-of-unity cases, and use these computations to define the Turaev-Viro invariants of closed 3-dimensional manifolds. Our approach is elementary and fairly self-contained. We develop the spin-networks from an algebraic point of view.

The Classical and Quantum
6j-symbols

1 Introduction

These notes grew out of a series of seminars held at the University of South Alabama during 1993 that were enhanced by regular e-mail among the three of us. We became interested in quantum diagrammatic representation theory following visits from Ruth Lawrence and Lou Kauffman to Mobile.

We develop the Clebsch-Gordan theory and the recoupling theory for representations of classical and quantum $U(sl(2))$ via the spin networks of Penrose [27] and Kauffman [16]. In these theories, the finite dimensional irreducible representations are realized in spaces of homogeneous polynomials in two variables. In the quantum case the variables commute up to a factor of q; $i.e.$ $yx = qxy$. The tensor product of two representations is decomposed as a direct sum of irreducibles, and the coefficients of the various weight vectors are computed explicitly. In the quantum case, when the parameter is a root of unity, we only decompose the representations modulo those that have trace 0.

We use the spin networks to develop the theory in the classical case for two reasons. First, they simplify and unify many of the tricky combinatorial facts. The simplification of the proofs is nowhere more apparent than in Theorem 2.7.14 where a plethora of identities is proven via diagram manipulations. Second, the spin networks are currently useful and quite popular in the quantum case (see for example [23], [18], [28]). One of our goals here is to explain the representation theory of quantum $sl(2)$ in the spin network framework. We know of no better explanation than to run through the classical case (which should be more familiar), and then to imitate the classical theory in the quantum case.

Here we give an overview. The set of (2 by 2) matrices of determinant 1 over the complex numbers forms a group called $SL(2)$. The finite dimensional irreducible representations of $SL(2)$ are well understood. In particular, it is known how to decompose the tensor product of two such representations into a direct sum of irreducibles. In this decomposition one can compute explicitly the image of weight vectors and such computations form the heart of the so-called Clebsch-Gordan theory. The finite dimensional representations of $SL(2)$ are the same as those of $U(sl(2))$ which is an algebra generated by symbols E, F and H subject to certain relations.

Furthermore, the tensor product of three representations can be decomposed in two natural ways. The comparison of these two decompositions is sometimes called *recoupling theory*, and the recoupling coefficients are known as the 6j-symbols. These symbols satisfy two fundamental identities (orthogonality and the Elliott-Biedenharn identity) that can be interpreted in terms of the decomposition of the union of two tetrahedra. In the Elliott-Biedenharn identity the tetrahedra are glued along a single face and recomposed as the union of three tetrahedra glued along an edge. For orthogonality the tetrahedra are glued along two faces, and the recomposition is not simplicial.

The symmetry of the 6j-symbols and their relationship to tetrahedra was for the most part a mystery, until Turaev and Viro [32] constructed 3-manifold invariants based on the analogous theory for quantum $sl(2)$. The identities satisfied by the 6j-symbols are also satisfied by their quantum analogues. The Elliott-Biedenharn identity corresponds to an Alexander [1] move on triangulations of a 3-manifold while the orthogonality condition can be interpreted as a Matveev [25] move on the dual 2-skeleton of a triangulation.

The Turaev-Viro invariants were based on work of Kirillov and Reshetikhin on the representation of quantum groups [19]. This work together with Reshetikhin-Turaev [29] formed a mathematically rigorous framework for the invariants of Witten [34]. Meanwhile Kauffman and Lins [18] gave a simple combinatoric approach to the invariants based on the Kauffman bracket and the spin networks of Penrose [27]. Piunikhin [28] showed that the Kauffman-Lins approach and the Turaev-Viro approach coincide.

Some of Kauffman's contributions to the subject can also be found in the papers [14], [15], and [17]. A more traditional algebraic approach to quantum groups can be found in [30]; in particular, they discuss from the outset the Hopf-algebra structures.

Lickorish's [23] definition of the Reshetikhin-Turaev invariants is of a combinatorial nature. The Kauffman-Lins [18] definition of the Turaev-Viro invariants is defined similarly. Neither of these combinatorial approaches relied on representation theory. However, the remarkable feature of quantum topology is that there are close connections between algebra and topology that were heretofore unimagined. The purpose of this paper is to explore these relations by examining the algebraic meaning of the diagrams and by using diagrams to prove algebraic results.

Here is our outline. Section 2 reviews the classical theory of representations of $U(sl(2))$. There is nothing new here, but we do show how the Clebsch-Gordan coefficients and the $6j$-symbols are computed in terms of the bracket expansion (at $A = 1$). In Section 3 we mimic these constructions to obtain the quantum Clebsch-Gordan and $6j$-symbols. In Section 4 we will define the quantum trace and discuss the recoupling theory in the root of unity case. Section 5 reviews the definitions of the Turaev-Viro invariants and proves that the definition is independent of the triangulation by means of the Pachner Theorem [26].

Acknowledgments. We all are grateful to L. Kauffman and R. Lawrence for the interesting conversations that we have had. Their visits to Mobile were supported by the University of South Alabama's Arts and Sciences Support and Development Fund. Additional financial support was obtained from Alabama EPSCoR for funding of travel for the first named author and support of a *Conference in Knot Theory, Low Dimensional Topology, and Quantum Groups* in Mobile in 1994. C. Pillen, B. Kuripta, K. Murasugi, and R. Peele provided us with valuable information. Jim Stasheff read a preliminary version of the text and provided us with many helpful comments. Cameron Gordon's past finanical support of Masahico Saito was greatly appreciated. Finally, we all gratefully acknowledge the support and patience that our wives have shown to us over the years.

2 Representations of $U(sl(2))$

2.1 Definition. Let **C** denote the complex numbers. The group $SL(2)$ is defined to be

$$SL(2) = \left\{ \begin{pmatrix} a & c \\ b & d \end{pmatrix} : a, b, c, d \in \mathbf{C}, \ ad - bc = 1 \right\}$$

where the law of composition is matrix multiplication. The associated Lie algebra $sl(2)$ consists of the set of matrices of trace 0:

$$sl(2) = \left\{ \begin{pmatrix} a & c \\ b & d \end{pmatrix} : a, b, c, d \in \mathbf{C}, \ a + d = 0 \right\}.$$

This is spanned by $E = \begin{pmatrix} 0 & 1 \\ 0 & 0 \end{pmatrix}$, $F = \begin{pmatrix} 0 & 0 \\ 1 & 0 \end{pmatrix}$, and $H = \begin{pmatrix} 1/2 & 0 \\ 0 & -1/2 \end{pmatrix}$. The *Lie bracket* is computed via $[A, B] = AB - BA$, so that $[E, F] = 2H$, $[H, E] = E$, and $[H, F] = -F$.

The Lie algebra $sl(2)$ is related to the Lie group $SL(2)$ via the exponential function, $\exp : sl(2) \rightarrow SL(2)$, which is defined by the power series:

$$\exp Q = \sum_{j=0}^{\infty} \frac{Q^j}{j!},$$

for $Q \in sl(2)$. The function exp maps a trace 0 matrix to a matrix with determinant 1.

2.2 Finite dimensional irreducible representations. The group $SL(2)$ acts on the vector space of linear combinations of variables x and y by

$$\begin{pmatrix} a & c \\ b & d \end{pmatrix} x = ax + by$$

and

$$\begin{pmatrix} a & c \\ b & d \end{pmatrix} y = cx + dy$$

where the action is extended linearly. This is called the *fundamental representation of $SL(2)$*.

More generally, define an action of $SL(2)$ on the space of polynomials in x and y by

$$\begin{pmatrix} a & c \\ b & d \end{pmatrix} x^r y^s = (ax + by)^r (cx + dy)^s.$$

One way to verify that this is a group action is to consider the embedding

$$x_1 \cdot \ \cdots \ \cdot x_{r+s} \mapsto \frac{1}{(r+s)!} \sum_{\sigma \in \Sigma_{r+s}} x_{\sigma 1} \otimes \cdots \otimes x_{\sigma(r+s)}$$

where for each k, x_k is either x or y and where the sum is over all permutations of $\{1, \ldots, r+s\}$. This embeds the space of homogeneous polynomials of degree $(r+s)$ in x and y into the tensor product of $(r+s)$ copies of the fundamental representation space. The tensor product $V \otimes W$ of representations V and W inherits an action via $g(v \otimes w) = gv \otimes gw$ where $v \in V$ and $w \in W$. Thus if V denotes the fundamental representation space, then $V^{\otimes(r+s)}$ is also a representation space. Furthermore, the image of the space of homogeneous polynomials consists of the subspace of tensors that are point-wise fixed under the action of the permutation group on the tensor factors of $V^{\otimes(r+s)}$, and this space is stable under the action of $SL(2)$.

It is customary to let V^j denote the set of homogeneous polynomials of degree $2j = r + s$ where $j \in \{0, 1/2, 1, 3/2, \ldots\}$. Note that $V^{1/2}$ is the fundamental representation, and V^0 is the trivial representation, $V^0 = \mathbf{C}$. The index j is sometimes called the *spin* of the representation V^j.

The associated Lie algebra $sl(2)$ acts on V^j as follows:

$$Ex^r y^s = \left.\frac{d}{dt}\right|_{t=0} \exp{(tE)}x^r y^s = sx^{r+1}y^{s-1},$$

$$Fx^r y^s = rx^{r-1}y^{s+1},$$

and

$$Hx^r y^s = \frac{r-s}{2}x^r y^s.$$

A *weight vector* is an eigenvector under the action of H in any representation; its eigenvalue is called its *weight*. For example $x^r y^s \in V^{(r+s)/2}$ is a weight vector of weight $\frac{r-s}{2}$. Observe that the set of weights in V^j is $\{j, j-1, j-2, \ldots, -j\}$, and by definition the corresponding weight vectors form a basis for V^j.

2.2.1 Well Known Theorem. (See [8] or [33], for example). *The representations of $SL(2)$ on V^j are irreducible.*

Proof. If W is an $SL(2)$-subrepresentation of V^j, then W is also invariant under the action of the algebra $sl(2)$ that is given above. Therefore, it is enough to show that the representation of $sl(2)$ on V^j is irreducible.

The matrix E acts by sending a weight vector to one of higher weight while F sends such to one of lower weight. Since the image of any non-zero vector under powers of E and F spans V^j, this representation is irreducible. $\square$

Remark. In the sequel, it will be more convenient to work with the universal enveloping algebra $U(sl(2))$. This is an algebra generated by symbols E, F, and H that are subject only to the relations $EF - FE = 2H$, $HE - EH = E$, and $HF - FH = -F$. The relations are motivated by the properties of the Lie bracket in $sl(2)$. A *representation* of either $sl(2)$ or of $U(sl(2))$ is determined by assigning to E, F, and H operators on a vector space that are subject to the relations above. The enveloping algebra acts by composition: $E^2 v = E(Ev)$ where v is a vector in the representation V.

In the discussion of Section 3, the representations V^j of $sl(2)$ and the enveloping algebra $U(sl(2))$ will have quantum analogues. There is a quantum analogue of the group $SL(2)$, but we will not use it to describe the representations.

2.2.2 Notation. The weight vector $x^r y^s$ should be indexed by the representation in which it lies and its weight. To this end let

$$e_{j,m} = x^{j+m} y^{j-m}.$$

The first subscript of e is the highest weight of the representation and indicates the dimension of the representation space $(\dim (V^j) = 2j + 1)$ while the second indicates the weight of the vector. Note that j and m are both half-integers and that $j + m$ and $j - m$ are integers. In this notation,

$$E e_{j,m} = (j - m)e_{j,m+1},$$

$$F e_{j,m} = (j + m)e_{j,m-1},$$

and

$$H e_{j,m} = m e_{j,m}.$$

2.2.3 Well Known Lemma.　　*(a) Let W denote a finite dimensional representation space for the algebra $U(sl(2))$. Let $v \in W$ denote a non-zero vector that satisfies:*

　　1. $Ev = 0$.

2. *The vector v has weight j.*

Then $j \in \{0, 1/2, 1, 3/2, \ldots\}$, and there is a unique linear map $\psi : V^j \to W$ such that $\psi(x^{2j}) = v$ and such that ψ commutes with the action of $U(sl(2))$.

(b) Every finite dimensional irreducible representation of $U(sl(2))$ is isomorphic to V^j for some $j \in \{0, 1/2, 1, 3/2, \ldots\}$.

Proof. Let $v_0 = v$, and for $r > 0$ let $v_r = F^r v$.

We assume by induction that v_r has weight $(j - r)$. Since $Hv = jv$ and $[H, F] = -F$, we have $Hv_{r+1} = HFv_r = -Fv_r + FHv_r = (j - (r + 1))v_{r+1}$; thus v_{r+1} has weight $(j - (r + 1))$.

Furthermore, we inductively assume that there are constants γ_r such that $Ev_r = \gamma_r v_{r-1}$. By using the relation $[E, F] = 2H$, we have that $Ev_{r+1} = EFv_r = 2Hv_r + FEv_r = (2(j - r) + \gamma_r)v_r$. Hence $\gamma_{r+1} = 2(j - r) + \gamma_r$; since $\gamma_0 = 0$, we have $\gamma_r = r(2j - r + 1)$.

Now $v_r = 0$ for some r because W is finite dimensional and $v_0, v_1, \ldots$ are eigenvectors for distinct eigenvalues of H. Suppose that $v_r = 0$ and $v_{r-1} \neq 0$. Then $Ev_r = 0 = r(2j - r + 1)v_{r-1}$. So $j = (r - 1)/2 \in \{0, 1/2, 1, 3/2, \ldots\}$, the subspace of W generated by v_0 is spanned by the linearly independent vectors $v_0, \ldots, v_{r-1}$, and this subspace is isomorphic to V^j. This proves (a).

(b) Let W denote a finite dimensional irreducible representation of $U(sl(2))$, and let w be a non-zero eigenvector of H. Let the integer r be such that $E^r w \neq 0$ while $E^{r+1} w = 0$. Then $v = E^r w$ satisfies the hypotheses of (a). Hence $W \approx V^j$ where $Hv = jv$. $\square$

2.2.4 Theorem. *Every finite dimensional representation space for $U(sl(2))$ decomposes as a direct sum of irreducible representations.*

Proof. See [33] or [8] for example. $\square$

2.3 Diagrammatics of $U(sl(2))$ invariant maps.   The Penrose spin networks facilitate the computation of $U(sl(2))$ invariant maps via diagrammatic techniques. At the heart of the networks are three elementary maps $\cup$, $\cap$, and $|$ that are defined in Section 2.3.1. Their relations are described in Lemma 2.3.2. The networks or *spin-nets* will consist of trivalent graphs embedded in the plane with non-negative half-integer labels on the edges. These labels will satisfy an admissibility condition at each vertex that will be made explicit as we continue the discussion.

2.3.1 Definition.   Consider the $U(sl(2))$ invariant maps $\cup$: $V^0 \to V^{1/2} \otimes V^{1/2}$, $\cap : V^{1/2} \otimes V^{1/2} \to V^0$, $\chi : V^{1/2} \otimes V^{1/2} \to V^{1/2} \otimes V^{1/2}$, and $| : V^{1/2} \to V^{1/2}$ that are defined on basis elements (and extended linearly) via

$$\cup(1) = i(x \otimes y - y \otimes x),$$

$$\cap(x \otimes x) = \cap(y \otimes y) = 0$$

while

$$\cap(x \otimes y) = i = -\cap(y \otimes x),$$

and

$$\chi(a \otimes b) = b \otimes a \quad \text{for} \quad a, b \in \{x, y\}$$

where $i = \sqrt{-1}$. Finally

$$|(a) = a \quad \text{for} \quad a \in \{x, y\}.$$

2.3.2 Lemma (Penrose [27]).

1. The maps $\cup$, $\cap$, χ, and $|$ commute with the action of $U(sl(2))$.

2. The fundamental binor identity holds:

$$\chi = |\otimes| + (\cup \circ \cap) : V^{1/2} \otimes V^{1/2} \to V^{1/2} \otimes V^{1/2} :$$

$$\chi = || + \overset{\smile}{\frown}$$

3. $(|\otimes \cap) \circ (\cup \otimes |) = | = (\cap \otimes |) \circ (| \otimes \cup) : V^{1/2} \rightarrow V^{1/2}$

under the identification of $\mathbf{C} \otimes V^{1/2} = V^{1/2} \otimes \mathbf{C} = V^{1/2}$.

$$\bigcup\!\bigcap = | = \bigcap\!\bigcup$$

4. $(\cap \otimes |) \circ (| \otimes \chi) = (| \otimes \cap) \circ (\chi \otimes |) : (V^{1/2})^{\otimes 3} \rightarrow V^{1/2}$

where as before we identify $\mathbf{C} \otimes V^{1/2} = V^{1/2} \otimes \mathbf{C} = V^{1/2}$.

$$\text{(diagram)} = \text{(diagram)}$$

5. $(\chi \otimes |) \circ (| \otimes \cup) = (| \otimes \chi) \circ (\cup \otimes |) : V^{1/2} \rightarrow (V^{1/2})^{\otimes 3}$.

$$\text{(diagram)} = \text{(diagram)}$$

6. $\cap \circ \cup : \mathbf{C} \rightarrow \mathbf{C}$

is multiplication by -2.

Proof. Items (1), (2), (3) and (6) are elementary computations. Item (4) is a general property that holds for any bilinear form $\cap$; similarly, item (5) follows for any "co-bilinear" form $\cup$. $\square$

2.3.3 Remarks.   Penrose and Kauffman introduced these maps in a diagrammatic context. The domain of a map represented by such a diagram appears at the bottom of the diagram, and vectors in the domain are fed through the diagram. Each critical level represents some map; the order of composition is $f \circ g(x) = f(g(x))$ and the diagram representing f appears above the diagram representing g. Thus, composition of the maps is indicated by vertical juxtaposition, and a vertical line denotes the identity mapping. When necessary, the identity is represented by a curve that has no critical points (with respect to the height function defined by the page: ↑). To say the least, the diagrammatic notation makes the identities in Lemma 2.3.2 memorable. They are illustrated in their diagrammatic form below each of the statements. The arcs in the diagrams will be referred to as *strings*. Here a diagrammatic proof that (5) follows from (3) and (4) is indicated.

Depending on the context in which multiplication takes place, we can represent $\cup$ and $\cap$ as a row vector: $(0, i, -i, 0)$, a column vector: $(0, i, -i, 0)^t$, or a matrix $\begin{pmatrix} 0 & i \\ -i & 0 \end{pmatrix}$. The beauty of the abstract tensor notation — where the joining of strings indicates contraction along the corresponding index and composition is obtained by vertical juxtaposition — is that the type of matrix composition is clearly represented by the diagram. Thus when no strings are joined the composition is an outer product; when two strings are joined at one end, this means a matrix product; and when two strings are joined to each other at both of their ends, we have an inner product.

Those readers familiar with the subject will recognize the fundamental binor identity as the Kauffman bracket evaluated at $A = 1$.

Since the map $\cup \circ \cap : V^{1/2} \otimes V^{1/2} \rightarrow V^{1/2} \otimes V^{1/2}$ is $U(sl(2))$ invariant, this map can be written as a linear combination of elements in the permutation group. The binor identity gives the linear relation explicitly. More generally, elements in the group algebra of Σ_n are precisely the set of operators that commute with the $U(sl(2))$ action [33] on $(V^{1/2})^{\otimes n}$. In the section that follows, we will define an algebra based on $\cup$ and $\cap$ (the Temperley-Lieb algebra) that allows us to find a basis for the $U(sl(2))$ invariant linear automorphisms of $(V^{1/2})^{\otimes n}$.

2.4 The Temperley-Lieb algebra. In order to decompose the tensor product of two irreducible representations as a direct sum of irreducibles, we will define maps that are linear combinations of compositions of tensor products of the $U(sl(2))$ invariant maps $\cup$, $\cap$ and $|$. One context in which many of the required maps are defined is the Temperley-Lieb algebra.

Let δ denote some element of a commutative ring R, and let $j \in \{0, 1/2, 1, 3/2, \ldots\}$. Consider the algebra $TL_{2j} = TL_{2j}(\delta)$ over R generated by the symbols $I, h_1, h_2, \ldots, h_{2j-1}$ that are subject to the relations:

1. $\qquad\qquad I^2 = I;$

2. $\qquad I h_k = h_k I = h_k \qquad\qquad$ for $k = 1, \ldots, 2j - 1;$

3. $\qquad\quad h_k h_\ell = h_\ell h_k \qquad\qquad$ for $|k - \ell| > 1;$

4. $\qquad\quad h_k h_k = \delta h_k \qquad\qquad$ for $k = 1, \ldots, 2j - 1;$

5. $\qquad h_k h_{k \pm 1} h_k = h_k \qquad\qquad$ for meaningful values of k.

Thus the set of monomials in $I, h_1, \ldots, h_{2j-1}$ spans the vector space TL_{2j} while the algebra structure is given by polynomial multiplication. There is an algebra isomorphism from TL_{2j} to an algebra of diagrams given by the map:

$$I \mapsto \underbrace{|\cdots|}_{2j}$$

and

$$h_k \mapsto \underbrace{|\cdots|}_{k-1} \; \substack{\cup \\ \cap} \; \underbrace{|\cdots|}_{2j-k-1}.$$

The *diagram algebra* consists of formal linear combinations of certain diagrams. The diagrams are generated by the diagrams representing I and h_k for $k = 1, \ldots, 2j-1$ that are indicated above. Any two such diagrams can be juxtaposed vertically to represent the product of two of the elements. Having been so juxtaposed, the product is rescaled vertically to fit into a standard size rectangle. Two diagrams that are isotopic via an isotopy that keeps the top and the bottom of the diagrams pointwise fixed represent the same element in the diagram algebra. For example, the product $h_2 h_1 h_3$ is depicted below. Always the element on the left of an expression is at the top of a diagram; thus the bottom most element corresponds to the inner most operator in a composition.

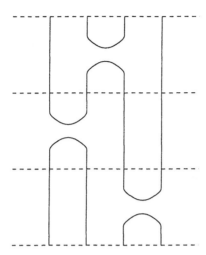

The key observation about the diagram algebra is that planar isotopies of arcs (that are properly and disjointly embedded in a rectangle) are generated by the topological moves that correspond to (a) the relationship depicted in Lemma 2.3.2 (3) and (b) interchanging distant critical points. For example, in the illustration above the critical points representing h_1 can be pushed down and those representing h_3 can be pushed up. Algebraically, this interchange represents the identity $h_1 h_3 = h_3 h_1$.

The correspondence between the Temperley-Lieb algebra and the algebra of diagrams shows that the dimension of TL_n is the nth *Catalan number*, $\frac{1}{n+1}\binom{2n}{n}$, where $n = 2j$. This result follows by establishing a one-to-one correspondence between the possible diagrams and the collection of legitimate arrangements of n pairs of parentheses (See also [16]).

Next we let the ground ring R denote the complex numbers, and choose matrix representations for the symbols $\cup$ and $\cap$.

2.4.1 Lemma. *Let j be a fixed element in $\{0, 1/2, 1, 3/2, \ldots\}$. For $\delta = -2$, there is a representation θ of TL_{2j} as an algebra of* *functions*

$$(V^{1/2})^{\otimes 2j} \to (V^{1/2})^{\otimes 2j}$$

The image $\theta(I)$ is the identity, and for $k = 1, \ldots, 2j-1$, the generator h_k is mapped to the composition $\cup \circ \cap$ where these are acting on the $(k, k+1)$ factors of the tensor product as in Section 2.3.1.

Proof. It is necessary to check that θ respects the defining relations (1-5) of the Temperley-Lieb algebra. The calculation follows from the diagrammatics of Lemma 2.3.2 part (3). $\square$

As we continue to discuss $U(sl(2))$, we will work with the Temperley-Lieb algebra under this representation without explicitly mentioning the map θ. Our justification for this notational abuse is given in Theorem 2.4.3.

2.4.2 Lemma. *Let $\delta = -2$. For any $j \in \{0, 1/2, 1, 3/2, \ldots\}$, there is a homomorphism ρ of the permutation group Σ_{2j} on $2j$ letters into TL_{2j} that is given by $\rho(\sigma_k) = I + h_k$ where σ_k is the transposition that interchanges k and $k+1$.*

Proof. Clearly, the images (under ρ) of distant transpositions commute. Furthermore,

$$\rho(\sigma_k \sigma_{k\pm1} \sigma_k) = I + h_k + h_{k\pm1} + h_k h_{k\pm1} + h_{k\pm1} h_k = \rho(\sigma_{k\pm1} \sigma_k \sigma_{k\pm1}).$$

Finally, $(I + h_k) \circ (I + h_k) = I + 2h_k + \delta h_k = I.\square$

Remark. The binor identity shows that the homomorphism ρ is a factor of the representation $\hat{\rho} : \Sigma_{2j} \to \text{Aut}((V^{1/2})^{\otimes 2j})$ where the permutation group acts on $(V^{1/2})^{\otimes 2j}$ by permuting tensor factors.

2.4.3 Theorem. *In case $\delta = -2$, the representation, θ of TL_{2j} on $(V^{1/2})^{\otimes 2j}$, is faithful for every $j \in \{0, 1/2, 1, 3/2, \ldots\}$.*

Proof. The proof depends on standard facts about the representation theory of $GL(2)$ and Σ_n. We refer the reader to the excellent text [31] for details. Let $n = 2j$. Since $\dim(TL_n) = \text{Catalan}(n)$ we must prove that $\dim(\theta(TL_n)) = \binom{2n}{n}/(n+1)$. The binor identity shows that $\theta(TL_n) = \hat{\rho}(\mathbf{C}[\Sigma_n])$, where $\hat{\rho}$ is the linear extension of the representation defined in the remark immediately above. We will establish that $\dim(\hat{\rho}(\mathbf{C}[\Sigma_n])) = \binom{2n}{n}/(n+1)$. The representation $\hat{\rho}$ is decomposed as in [33]. Namely, as Σ_n representations,

$$(V^{1/2})^{\otimes n} = \oplus_T (W_T)^{d(T)}$$

where:

(1) the index T ranges over all 2-row Young frames with n boxes (*i.e.* $T = (r, s), n = r + s$, and $0 \le r \le s$);

(2) the summand W_T is the irreducible representation of Σ_n that corresponds to the Young frame T;

(3) the exponent $d(T)$ is a positive integer that, incidentally, is equal to the dimension of the representation of $GL(2)$ corresponding to the Young frame T.

Now $\mathbf{C}[\Sigma_n]$ is a semi-simple algebra because the group Σ_n is finite. The Wedderburn theory of semi-simple algebras [9], applied to $\mathbf{C}[\Sigma_n]$, asserts that as algebras

$$\hat{\rho}(\mathbf{C}[\Sigma_n]) \approx \oplus_T \text{Mat}(n_T \times n_T)$$

where $n_T = \dim W_T$ and $\text{Mat}(\cdot \times \cdot)$ denotes the algebra of square matrices. It follows that

$$\dim(\hat{\rho}(\mathbf{C}[\Sigma_n])) = \sum_T n_T^2.$$

For $T = (r, n - r)$ with $0 \leq r \leq \lfloor n/2 \rfloor$ (where $\lfloor \cdot \rfloor$ denotes the greatest integer function), we have

$$n_T = \frac{n + 1 - 2r}{n + 1 - r} \binom{n}{r}.$$

This is the number of ways of filling in the n boxes in the Young frame T with the integers $1, 2, \ldots, n$ in such a way that numbers increase across both rows and increase down all columns. The Young frame has $n - r$ boxes on the top row and r boxes on the bottom row.

The proof will follow from the following interesting combinatorial identity for Catalan numbers:

$$\frac{1}{n + 1} \binom{2n}{n} = \sum_{r=0}^{\lfloor n/2 \rfloor} \left[\frac{n + 1 - 2r}{n + 1 - r} \binom{n}{r} \right]^2.$$

Let $G(r, n - r) = n_{(r, n-r)}$ denote the number of legitimate fillings of the Young frame with r boxes on the bottom row and $n - r$ boxes on the top. For a two row rectangular array,

$$G(r, r) = \frac{1}{r + 1} \binom{2r}{r},$$

so in fact $G(r, r)$ is the rth Catalan number. We wish to show that

$$G(n, n) = \sum_{s=0}^{\lfloor n/2 \rfloor} [G(s, n - s)]^2.$$

Each term in the sum on the right is the square of the number of ways of filling in a smaller Young frame. For $s \leq \lfloor n/2 \rfloor$, we consider the Young frame (n, n) to be decomposed as the union of a frame $(s, n - s)$ and its mirror image. For example,

A filling of the frame $(s, n - s)$ with the integers $1, \ldots, n$ together with a filling of its mirror image with the integers $n + 1, \ldots, 2n$ yields a filling of the rectangular frame. Therefore, the sum on the right is no larger than $G(n, n)$.

On the other hand, let a filling of the rectangular frame be given. Then consider the subset of the rectangular array that contains the numbers $1, \ldots, n$. This subset is convex and forms a smaller frame of type $(s, n - s)$. Thus we have a filling of it and a filling of its mirror image. Therefore, $G(n, n)$ is no larger than the sum on the right. This proves the combinatorial identity. Consequently, the representation is faithful. $\square$

2.5 Tensor products of irreducible representations. Recall that if V and W are spaces on which the group $SL(2)$ acts, then there is an action given on the tensor product by $g(v \otimes w) = gv \otimes gw$. An element X in the associated Lie algebra, $sl(2)$, acts on tensor products via the Leibniz rule, $X(v \otimes w) = X(v) \otimes w + v \otimes X(w)$ since the action is determined by differentiation. Notice that if v and w are weight vectors, then so is $v \otimes w$, and its weight is the sum of the weights of v and w. Recall that V^j is isomorphic to a sub-representation of the $2j$-fold tensor product of the fundamental representation via the map

$$\phi_j : x_1 \cdots \cdots x_{2j} \mapsto \frac{1}{(2j)!} \sum_\sigma x_{\sigma(1)} \otimes \cdots \otimes x_{\sigma(2j)}$$

the sum being taken over all permutations of $\{1, \ldots, 2j\}$ and $x_i \in \{x, y\}$.

2.5.1 The projectors. The projection of $(V^{1/2})^{\otimes 2j}$ onto the image $\phi_j(V^j)$ can be written in terms of the Temperley-Lieb elements as the map

$$+_{2j} = \frac{1}{(2j)!} \sum_{\sigma \in \Sigma_{2j}} \rho(\sigma).$$

Observe that $+_{2j} \circ +_{2j} = +_{2j}$ so that this map is indeed a projection.

For example,

$$+_3 = I_3 + \frac{2}{3}(h_1 + h_2) + \frac{1}{3}(h_1 h_2 + h_2 h_1)$$

$$= |\,|\,| + \frac{2}{3}\left[\,\bigcap^\cup \otimes \Big| + \Big| \otimes \bigcap^\cup \, \right]$$

$$+ \frac{1}{3}\left[\left(\bigcap^\cup \otimes \Big| \right) \circ \left(\Big| \otimes \bigcap^\cup \right) + \left(\Big| \otimes \bigcap^\cup \right) \circ \left(\bigcap^\cup \otimes \Big| \right) \right].$$

In section 3.5, the quantum analogues of these projectors are defined. In the quantum case, they are key to developing the entire theory.

2.5.2 Definition. Let $\overset{1}{\cup} = \cup : \mathbf{C} \to V^{1/2} \otimes V^{1/2}$. Having defined $\overset{n-1}{\cup} : \mathbf{C} \to (V^{1/2})^{\otimes 2(n-1)}$, define $\overset{n}{\cup}$ to be the composition

$$\overset{n}{\cup} : \mathbf{C} \overset{\overset{n-1}{\cup}}{\to} \underbrace{V^{1/2} \otimes \ldots \otimes V^{1/2}}_{2(n-1)} \overset{=}{\to}$$

$$\underbrace{V^{1/2} \otimes \ldots \otimes V^{1/2}}_{(n-1)} \otimes \mathbf{C} \otimes \underbrace{V^{1/2} \otimes \ldots \otimes V^{1/2}}_{(n-1)}$$

$$\overset{1 \otimes \cup \otimes 1}{\to} (V^{1/2})^{\otimes 2n}.$$

The map $\overset{n}{\cap}$ is defined dually.

Let $S \subset \{1, \ldots, n\}$. Define

$$x^S_m = \begin{cases} y & \text{if } m \in S \\ x & \text{if } m \notin S \end{cases},$$

and define

$$\bar{x}_m^S = \begin{cases} x & \text{if } m \in S \\ y & \text{if } m \notin S \end{cases}$$

2.5.3 Lemma.

$$\overset{n}{\cup}(1) = i^n \sum_{S \subset \{1,\dots,n\}} (-1)^{|S|} x_1^S \otimes \cdots \otimes x_n^S \otimes \bar{x}_n^S \otimes \cdots \otimes \bar{x}_1^S.$$

Proof. The proof follows by induction. $\square$

2.5.4 Definition. Suppose that $a, b \in \{0, 1/2, 1, 3/2, \dots\}$. Let $j \in \{a + b, a + b - 1, \dots, |a - b| + 1, |a - b|\}$. Such a triple of half-integers (a, b, j) is said to be *admissible*. Notice that admissibility is a symmetric condition in a, b, and j. Define an $U(sl(2))$ invariant map

$$\overset{ab}{\underset{j}{Y}} : (V^{1/2})^{\otimes 2j} \rightarrow (V^{1/2})^{\otimes 2a} \otimes (V^{1/2})^{\otimes 2b}$$

as follows.

$$\overset{ab}{\underset{j}{Y}} = (\overset{+}{}_{2a} \otimes \overset{+}{}_{2b}) \circ (\,|_{a+j-b} \otimes \overset{a+b-j}{\cup} \otimes \,|_{b+j-a}\,) \circ \overset{+}{}_{2j}$$

where $|_m$ is the identity map on the m-fold tensor power of $V^{1/2}$.

The map $(\mu_a \otimes \mu_b) \circ \overset{ab}{\underset{j}{Y}} \circ \phi_j$ —where ϕ_ℓ is the isomorphism of V^ℓ with the symmetric tensors while $\mu_\ell(x_1 \otimes \cdots \otimes x_{2\ell}) = x_1 \cdot \cdots \cdot x_{2\ell}$ for $\ell = j, a$, or b — is called the *Clebsch-Gordan map*: $V^j \rightarrow V^a \otimes V^b$.

2.5.5 Theorem. *There is a direct sum decomposition*

$$V^a \otimes V^b = \bigoplus_j \mu_a \otimes \mu_b \left(\overset{ab}{\underset{j}{Y}} \left(\phi_j \left(V^j \right) \right) \right)$$

where the sum is taken over all j such that (a, b, j) is admissible. Furthermore, if (a, b, j) is an admissible triple, then any $U(sl(2))$ invariant map $V^j \to V^a \otimes V^b$ is a scalar multiple of $\mu_a \otimes \mu_b \circ \underset{j}{\overset{ab}{Y}} \circ \phi_j$. Finally,

$$\mu_a \otimes \mu_b \left(\underset{j}{\overset{ab}{Y}} \left(\phi_j \left(x^{2j} \right) \right) \right) =$$

$$i^{a+b-j} \sum_{L \subset \{1, \ldots, a+b-j\}} (-1)^{|L|} (x^{2a-|L|} y^{|L|}) \otimes (x^{|L|+b+j-a} y^{a+b-j-|L|})$$

$$= i^{a+b-j} \sum_{k=0}^{a+b-j} (-1)^k \binom{a+b-j}{k} x^{2a-k} y^k \otimes x^{b-a+j+k} y^{a+b-j-k}.$$

Proof. The map $\underset{j}{\overset{ab}{Y}}$ is $U(sl(2))$ invariant because it is the composition of $U(sl(2))$ invariant maps $\overset{}{+}$ and $\overset{n}{\cup}$. The formula for $\left(\mu_a \otimes \mu_b \left(\underset{j}{\overset{ab}{Y}} \left(\phi_j \left(x^{2j} \right) \right) \right) \right)$ follows by computation using Lemma 2.5.3; thus $\underset{j}{\overset{ab}{Y}} \neq 0$ for (a, b, j) admissible. The tensor product $V^a \otimes V^b$ has dimension $(2a+1)(2b+1)$ while the image of V^j has dimension $(2j+1)$. Since

$$\sum_j (2j+1) = (2a+1)(2b+1)$$

where the sum is taken over $j \in \{a+b, a+b-1, \ldots, |a-b|\}$, the images $\mu_a \otimes \mu_b \left(\underset{j}{\overset{ab}{Y}} \left(\phi_j \left(V^j \right) \right) \right)$ span $V^a \otimes V^b$. Thus $\mu_a \otimes \mu_b \left(\underset{j}{\overset{ab}{Y}} \left(\phi_j \left(V^j \right) \right) \right)$ is the only subspace of $V^a \otimes V^b$ that is isomorphic to V^j. Consequently every $U(sl(2))$ invariant map $V^j \to V^a \otimes V^b$ must be a multiple of this map. $\square$

Example. Consider $V^{1/2} \otimes V^{1/2}$. According to Theorem 2.5.5, this tensor product decomposes as the direct sum of V^0 and V^1. The map $\curlyvee_0^{1/2,1/2}$ coincides with $\cup$ while

$$\curlyvee_1^{1/2,1/2} (\phi x^2) = x \otimes x,$$

$$\curlyvee_1^{1/2,1/2} (\phi y^2) = y \otimes y,$$

and finally

$$\curlyvee_1^{1/2,1/2} (\phi xy) = 1/2(x \otimes y + y \otimes x).$$

2.5.6 The Clebsch-Gordan coefficients. Let $e_{j,t}$ denote the weight vector $x^{j+t}y^{j-t}$ in V^j of weight t. We have maps $\mu_a \otimes \mu_b \circ \curlyvee_j^{ab} \circ \phi_j : V^j \to V^a \otimes V^b$ provided that (a,b,j) are admissible. Define the *Clebsch-Gordan coefficient* $C_{u,v,t}^{a,b,j}$ to be the coefficient in the sum

$$\mu_a \otimes \mu_b \left(\curlyvee_j^{ab} (\phi_j(e_{j,t})) \right) = \sum_{u+v=t} C_{u,v,t}^{a,b,j} e_{a,u} \otimes e_{b,v}.$$

2.5.7 Lemma. *The Clebsch-Gordan coefficients satisfy the following recursion relation*

$$(j+t+1)C_{u,v,t}^{a,b,j} = (a+u+1)C_{u+1,v,t+1}^{a,b,j} + (b+v+1)C_{u,v+1,t+1}^{a,b,j}.$$

Hence,

$$C_{u,v,t}^{a,b,j} = i^{a+b-j} \frac{(j+t)!(j-t)!(a+b-j)!}{(2j)!(a+u)!(b+v)!}.$$

$$\sum_{z,w:\ z+w=j-t} (-1)^{a-u+z} \frac{(a+u+z)!(b+v+w)!}{z!w!(a-u-z)!(b-v-w)!}.$$

Proof. The recursion relation is found by applying F to both sides of the equation that defines the Clebsch-Gordan coefficient. The closed form is determined by solving the recursion using the value

$$C_{u,v,j}^{a,b,j} = i^{a+b-j}(-1)^{a-u}\frac{(a+b-j)!}{(a-u)!(b-v)!}$$

that is given in 2.5.5. $\square$

2.5.8 Diagrammatics for weight vectors. By definition of ϕ, the weight vector $\phi_t(e_{j,t})$ is the image under $+_{2j}$ of

$$\underbrace{x\otimes\cdots\otimes x}_{m}\otimes\underbrace{y\otimes\cdots\otimes y}_{n}$$

where $m = j + t$, $n = j - t$.

We represent x (resp. y) by a white (resp. black) vertex with a string coming out from the top: $\text{\char"F6}$ (resp. $\text{\char"F6}$). Then the weight vector $\phi_t(e_{j,t})$ in the image of V^j is represented by:

These conventions have been known to physicists (see [16]).

It is convenient to introduce similar diagrams for dual vectors. Consider the dual vector space $(V^{1/2})^*$. We represent the dual basis vectors x^* and y^* of this dual space diagrammatically by $\text{\char"F6}$ and $\text{\char"F6}$, respectively. For parallel strings representing tensor products of the fundamental representation, putting one of these dots on the top of the strings algebraically means that we take the values of the pairing among vectors and dual vectors. In particular, $\text{\char"F6}$ means the pairing between x and x^* which is equal to one. We have, then, that the dual to

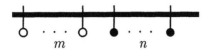

is the vector

$$\frac{(m+n)!}{m!\ n!}$$

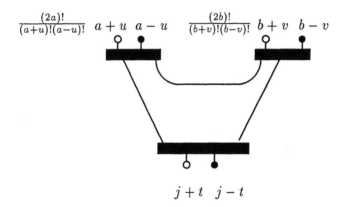

The Clebsch-Gordan coefficient $C_{u,v,t}^{a,b,j}$ is given by the following diagram.

From the closed formula given in Lemma 2.5.7 for the Clebsch-Gordan coefficients, one cannot easily see the symmetry properties of the coefficients under replacing $e_{j,m}$ with $e_{j,-m}$. However, this symmetry and the probabilistic nature of the coefficients is more apparent in the network evaluation.

2.6 The $6j$-symbols.   Here we consider the space of $U(sl(2))$ invariant maps $V^k \to V^a \otimes V^b \otimes V^c$. We will construct such maps in two different ways. First, consider the composition

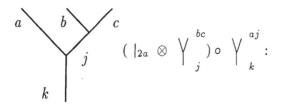

$$(V^{1/2})^{\otimes 2k} \to (V^{1/2})^{\otimes 2a} \otimes (V^{1/2})^{\otimes 2b} \otimes (V^{1/2})^{\otimes 2c}$$

for various values of j. Second, consider the composition

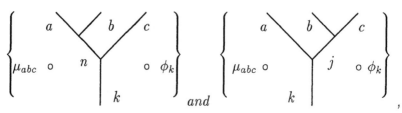

for various values of n.

The values of j and n are restricted so that (b, c, j), (a, j, k), (a, b, n), and (n, c, k) all form admissible triples. Alternatively, if one of these triples is not admissible, then we may declare the corresponding map Y to be the zero map. (Recall that if (a, b, j) is admissible, then so are the triples obtained by permuting a, b, and j.)

2.6.1 Lemma. *The sets*

$$\left\{ \mu_{abc} \circ \quad\quad \circ \phi_k \right\} \quad and \quad \left\{ \mu_{abc} \circ \quad\quad \circ \phi_k \right\},$$

as the indices j and n range in such a way that (b, c, j), (a, j, k),

(a, b, n), and (n, c, k) *form admissible triples, form bases for the vector space of $U(sl(2))$ invariant linear maps $V^k \to V^a \otimes V^b \otimes V^c$. Here $\mu_{abc} = \mu_a \otimes \mu_b \otimes \mu_c$ is the tensor product of the multiplication maps and $\phi_j : V^j \to (V^{1/2})^{\otimes 2j}$ sends a a homogeneous polynomial in x and y to a symmetric tensor.*

Proof. The triple tensor product $V^a \otimes V^b \otimes V^c$ decomposes as $(\oplus_n V^n) \otimes V^c = \oplus_n (V^n \otimes V^c)$ where the direct sum is taken over all n such that (a, b, n) is admissible, by Theorem 2.5.5. For each such n, $(V^n \otimes V^c)$ contains at most one copy of V^k. Thus $\hom_{U(sl(2))}(V^k, V^a \otimes V^b \otimes V^c)$ decomposes as a direct sum of the 1-dimensional spaces $\hom_{U(sl(2))}(V^k, V^n \otimes V^c)$. Similarly, it decomposes as a direct sum of the 1-dimensional spaces $\hom_{U(sl(2))}(V^k, V^a \otimes V^j)$. $\square$

2.6.2 Definition. Define the *6j-symbol* to be the coefficient $\left\{ \begin{array}{ccc} a & b & n \\ c & k & j \end{array} \right\}$ in the following equation.

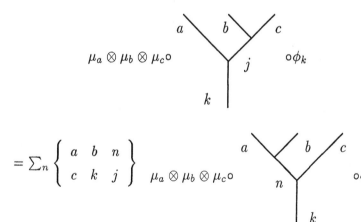

By convention, $\left\{ \begin{array}{ccc} a & b & n \\ c & k & j \end{array} \right\} = 0$ if any of the triples (b, c, j), (a, j, k), (a, b, n), (n, c, k) is not admissible.

In the spaces $\hom_{U(sl(2))}(V^k, V^a \otimes V^b \otimes V^c)$, we have the two bases that are defined by these trees, and the $6j$-symbol is the change of basis matrix. For example, consider the case when $a = b = c = k = 1/2$. One can compute directly from the definitions that the possible values for j and n are 0 and 1, and that:

$$\left\{ \begin{matrix} 1/2 & 1/2 & 0 \\ 1/2 & 1/2 & 0 \end{matrix} \right\} = -1/2,$$

$$\left\{ \begin{matrix} 1/2 & 1/2 & 1 \\ 1/2 & 1/2 & 0 \end{matrix} \right\} = 1,$$

$$\left\{ \begin{matrix} 1/2 & 1/2 & 0 \\ 1/2 & 1/2 & 1 \end{matrix} \right\} = 3/4,$$

and

$$\left\{ \begin{matrix} 1/2 & 1/2 & 1 \\ 1/2 & 1/2 & 1 \end{matrix} \right\} = 1/2.$$

In Section 2.8, we give a recursive method to compute the values of the $6j$-symbols in general.

Before we state and prove the main results about the $6j$-symbols we need to define an $U(sl(2))$ invariant map $\bigwedge_{ab}^{j}$: $(V^{1/2})^{\otimes 2a} \otimes (V^{1/2})^{\otimes 2b} \to (V^{1/2})^{\otimes 2j}$, for admissible triples (a, b, j) as follows:

$$\bigwedge_{ab}^{j} = \textbf{+}_{2j} \circ \left(|_{a+j-b} \otimes \overset{a+b-j}{\cap} \otimes |_{j+b-a} \right) \circ \left(\textbf{+}_{2a} \otimes \textbf{+}_{2b} \right).$$

The composition $\mu_j \circ \bigwedge_{ab}^{j} \circ (\phi_a \otimes \phi_b) : V^a \otimes V^b \to V^j$ is also $U(sl(2))$ invariant.

2.6.3 Diagrammatic computations. The elegance of the spin-net notation that we have developed so far is that it facilitates otherwise tedious calculations. There is a slight disadvantage in that all the calculations are performed in the tensor power,

$(V^{1/2})^{\otimes 2j}$, of the fundamental representation rather than on the irreducible representations, V^j, themselves. This is a small price to pay since we can always compose the maps defined via spin-nets with ϕs and μs.

Recall that a *spin-net* is an embedding in the plane of a graph with edges labeled by non-negative half-integers in which each vertex has valence 3. The three edges coincident at a vertex must form an admissible triple. The half-integer labels represent the spin carried by an edge. When we need to emphasize the number of strings represented by an edge (and hence the number of tensor factors of the fundamental representation carried by an edge), we will label the edges with natural numbers that are twice the half-integers. The notation suffers from this minor inconsistency, but we have found that the meaning of the labels is clear within the context in which it is written.

In addition, we assume that the embedding is in general position with respect to a fixed height function: Thus each vertex appears at a distinct level, the critical points on each edge are non-degenerate, and these critical points are at distinct levels from the vertices. Furthermore, some edges may be marked with symmetrizers: $+$. More precisely, we include valence 2 vertices in which the two incoming edges have the same label.

The principal results of the current section are the orthogonality and the Elliott-Biedenharn identities that are satisfied by the $6j$-symbols. We will give proofs of these relationships (and others) that are simply manipulations of diagrams. To this end we state a diagrammatic lemma.

2.6.4 Lemma.   *The following relationships hold among the $U(sl(2))$ invariant maps* $\bigwedge$, $\bigvee$, $\overset{n}{\cup}$, $|_n$, *and* $\overset{n}{\cap}$. *(Here we identify* $\mathbf{C} \otimes V$ *and* V *for any* $U(sl(2))$ *space* V.)

1. $\left(|_n \otimes \overset{n}{\cap} \right) \circ \left(\overset{n}{\cup} \otimes |_n \right) = |_n = \left(\overset{n}{\cap} \otimes |_n \right) \circ \left(|_n \otimes \overset{n}{\cup} \right).$

$$\bigcup_n \; = \; |_n \; = \; \bigcap_n$$

2. $\left(|_n \otimes +_n \right) \circ \overset{n}{\cup} = \left(+_n \otimes |_n \right) \circ \overset{n}{\cup}.$

$$\bigcup_n \!\! \vdash \; = \; \dashv \! \bigcup_n$$

3. $\overset{n}{\cap} \circ \left(|_n \otimes +_n \right) = \overset{n}{\cap} \circ \left(+_n \otimes |_n \right)$

$$\bigcap_n \!\! \vdash \; = \; \dashv \! \bigcap_n$$

4.

$$\left(\bigwedge\nolimits_{jb}^{a} \otimes \; |_{2b} \right) \circ \left(|_{2j} \otimes \overset{2b}{\cup} \right) = \bigvee\nolimits_{j}^{ab}$$

$$= \left(|_{2a} \otimes \bigwedge\nolimits_{aj}^{b} \right) \circ \left(\overset{2a}{\cup} \otimes |_{2j} \right).$$

$$\bigwedge\!\!\bigcup \; = \; \bigvee \; = \; \bigcup\!\!\bigwedge$$

5.

$$\left(\Big|_{2j} \otimes \overset{2b}{\cap} \right) \circ \left(\overset{jb}{\underset{a}{Y}} \otimes \Big|_{2b} \right) = \overset{j}{\underset{ab}{\lambda}}$$

$$= \left(\overset{2a}{\cap} \otimes \Big|_{2j} \right) \circ \left(\Big|_{2a} \otimes \overset{aj}{\underset{b}{Y}} \right).$$

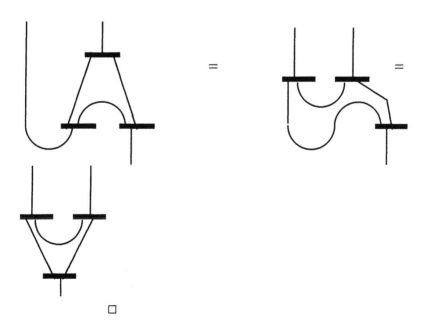

Proof. Part 1 follows from Lemma 2.3.2 part 3, and induction since the cancellation of a $\cup$ with a $\cap$ can occur regardless of the tensor factors on which those maps are acting. Part 2 follows by induction and part 5 of Lemma 2.3.2. Part 3 follows similarly using part 4 of Lemma 2.3.2. Part 4 and part 5 are proved using the previous results as follows.

□

2.6.5 Remark. Consider the collection of proper embeddings of two and three valent graphs in a rectangle whose edges are parallel to the coordinate axes in the plane. The free end points of the edges of the graph are embedded in the top and bottom edges of the rectangle. If two such embedded graphs are isotopic via an isotopy that keeps the boundary fixed, then there is an isotopy between them that can be decomposed as a sequence of moves that are the diagrammatic descriptions of items 1 through 5 in Lemma 2.6.4. The valence two vertices are represented in the Lemma by the projectors, and the valence three vertices are represented by the Clebsh-Gordan maps.

To find such a nice isotopy, one replaces a given isotopy by one that is in general position with respect to the height function defined on the rectangle. The existence of the generic isotopy is guaranteed by a transversality argument, and a similar transversality argument decomposes the nice isotopy into a finite number of pieces each of which is of the diagrammatic form specified.

2.6.6 Theorem (Orthogonality). *Suppose that* (a, b, n), $(c, k, n), (a, b, m)$, *and* (c, k, m) *are admissible triples. Then then* 6j-*symbols satisfy the following relation:*

$$\sum_j \left\{ \begin{array}{ccc} b & c & j \\ k & a & n \end{array} \right\} \left\{ \begin{array}{ccc} a & b & m \\ c & k & j \end{array} \right\} = \delta_{m,n}.$$

Proof. Define

$$
\underset{s}{\overset{m}{\diagdown}} r \underset{t}{\overset{p}{\diagup}} = \left(\mid_{2m} \otimes \underset{rt}{\overset{p}{\bigwedge}} \right) \circ \left(\overset{mr}{\underset{s}{\bigvee}} \otimes \mid_{2t} \right)
$$

$$
= \left(\underset{sr}{\overset{m}{\bigwedge}} \otimes \mid_{2p} \right) \circ \left(\mid_{2s} \otimes \overset{rp}{\underset{t}{\bigvee}} \right),
$$

and define

$$\overset{m\quad p}{\underset{s\quad t}{\bigtimes}}\!{}_{u} \;=\; \overset{mp}{\underset{u}{\mathsf{Y}}} \;\circ\; \underset{st}{\overset{u}{\bigwedge}}.$$

Then we have the following calculation.

$$= \sum_u \left\{ \begin{matrix} m & p & u \\ t & s & r \end{matrix} \right\}$$

Therefore,

$$\overset{m\quad p}{\underset{s\quad t}{\bigtimes}}\!{}_r \;=\; \sum_u \left\{ \begin{matrix} m & p & u \\ t & s & r \end{matrix} \right\} \overset{m\quad p}{\underset{s\quad t}{\bigtimes}}\!{}_u.$$

This last equation is called the *recoupling formula*. Although we are making calculations here on the tensor powers of the fundamental representations rather than on the irreducible representations, the same coefficients arise: Each of the maps in question begins and ends in a projector, and so the domain and range of the given maps can be chosen to be the isomorphic images under ϕ_j for various values of j of the irreducible representations.

In the same spirit we have the following:

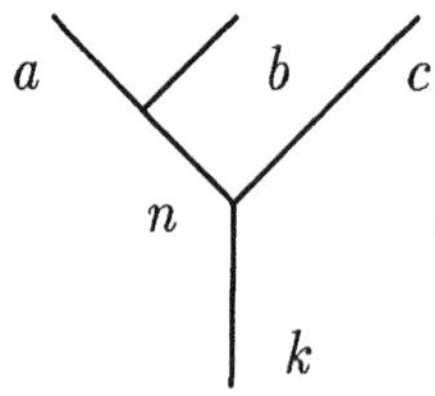

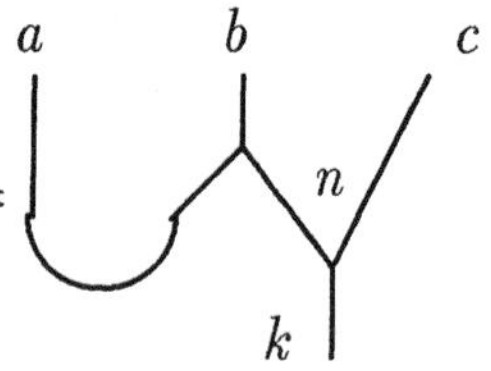

$$= \sum_j \begin{Bmatrix} b & c & j \\ k & a & n \end{Bmatrix}$$

$$= \sum_j \begin{Bmatrix} b & c & j \\ k & a & n \end{Bmatrix}$$

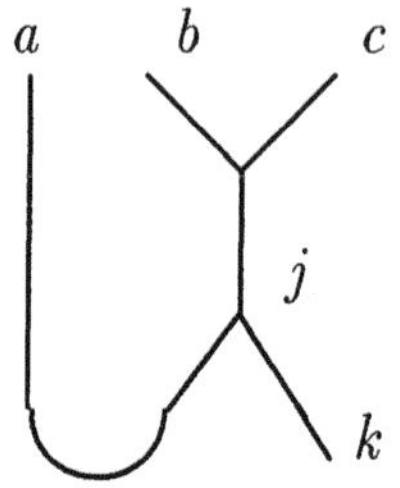

$$= \sum_m \sum_j \begin{Bmatrix} b & c & j \\ k & a & n \end{Bmatrix} \begin{Bmatrix} a & b & m \\ c & k & j \end{Bmatrix}$$

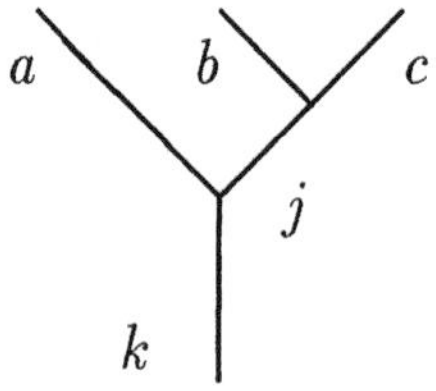

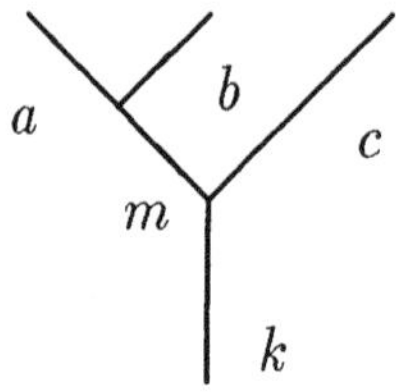

Therefore,

$$\sum_j \left\{ \begin{matrix} b & c & j \\ k & a & n \end{matrix} \right\} \left\{ \begin{matrix} a & b & m \\ c & k & j \end{matrix} \right\} = \delta_{m,n}$$

because the set 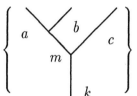 as m varies is linearly independent.

This completes the proof. □

2.6.7 Theorem (Elliott-Biedenharn Identity). *If (g,e,h) is an admissible triple, then the following relation holds among the $6j$-symbols.*

$$\left\{ \begin{matrix} c & d & h \\ g & e & f \end{matrix} \right\} \cdot \left\{ \begin{matrix} b & h & k \\ g & a & e \end{matrix} \right\}$$

$$= \sum_j \left\{ \begin{matrix} b & c & j \\ f & a & e \end{matrix} \right\} \cdot \left\{ \begin{matrix} j & d & k \\ g & a & f \end{matrix} \right\} \cdot \left\{ \begin{matrix} c & d & h \\ k & b & j \end{matrix} \right\}.$$

Proof. We consider the rooted trees that have four branchs depicted below as maps $V^g \to V^a \otimes V^b \otimes V^c \otimes V^d$. These are defined in terms of the maps $\vee$, ϕ_g, and μ_a, ..., μ_d in an obvious fashion. As the branches move, $6j$ symbols are engendered as indicated in the diagram. The proof follows from the picture below because the family of maps that appear lowest on the diagram as k and h vary is linearly independent.

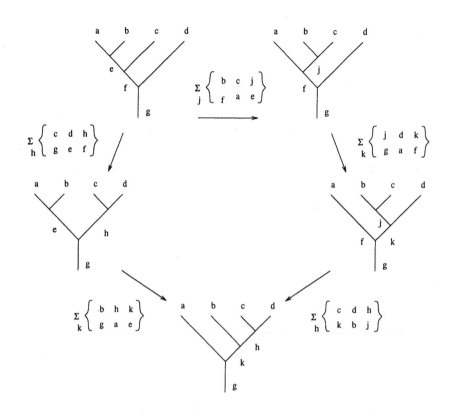

2.6.8 Associating the $6j$-symbol to the dual skeleton of the tetrahedron.
Consider the 2-dimensional cell complex that occurs when the branch labeled by b is moved from the

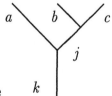

right side of the tree   to the left side of the tree

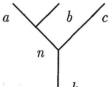

. This complex is depicted in Figure 1. Each face of the 2-dimensional complex arises as the cartesian product of an edge of the tree and a unit interval; the interval factor is thought of as a time parameter in the deformation between the two trees. The vertex of the complex occurs as the edge labeled b passes through the lower junction of three edges. The 2-dimensional complex has 1 vertex, 4 edges, and 6 faces; thus it is the dual complex to a tetrahedron as indicated in the figure.

Let us associate to this complex a $6j$-symbol. Then consider the orthogonality conditions and the Elliott-Biedenharn identities. Either side of each equation can be similarly thought of as a 2-dimensional cell complex. For example, in Figure 2 one side of the orthogonality relation is depicted both as a movie description and as a 2-dimensional complex; the 2-complex is the "time elapsed" version of the accompanying movie. The sucessive stills of the movie differ by a $6j$-symbol and these symbols are associated to the vertices of the 2-dimensional complex on the right of the figure. The trees in the stills are the trees that represent the maps in the orthogonality relation. Figure 3 depicts the other side of the orthogonality relation, and the 2-dimensional complex here differs from the previous one by one of the Matveev moves [25, 16, 32]. Similarly, Figure 4 depicts in a movie fashion the three trees that appear on one side of the Elliott-Biedenharn identity. The sucessive stills in the movie differ by a $6j$-symbol. In Figure 5 the stills in the movie are the trees on the other side of the Elliott-Biedenharn identity. The 2-complex in either figure represents the

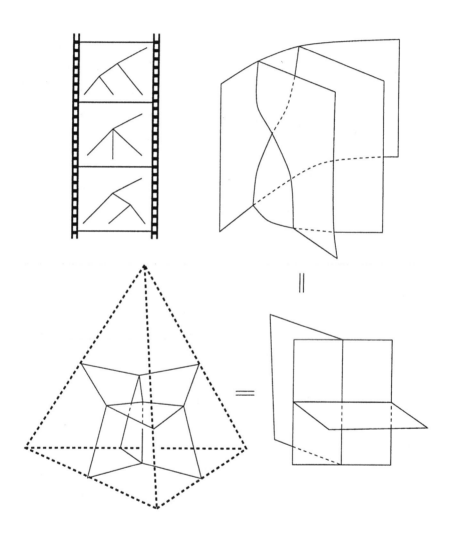

Figure 1: A movie of a 6*j*-symbol and the Matveev complex

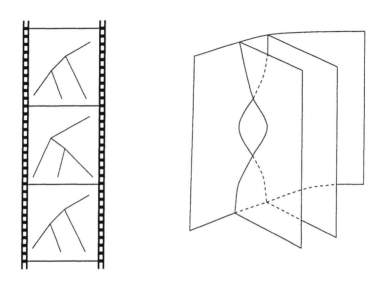

Figure 2: A movie of $6j$-symbols and orthogonality (left hand side)

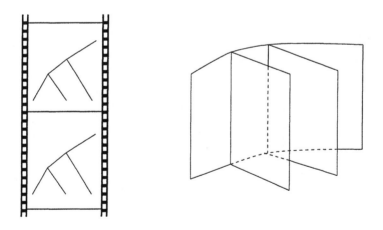

Figure 3: A movie of $6j$-symbols and orthogonality (right hand side)

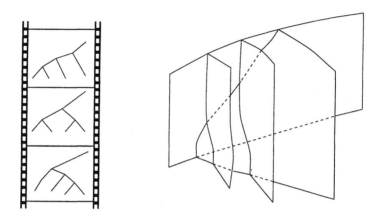

Figure 4: A movie of 6j-symbols and the Elliott-Biedenharn identity (left hand side)

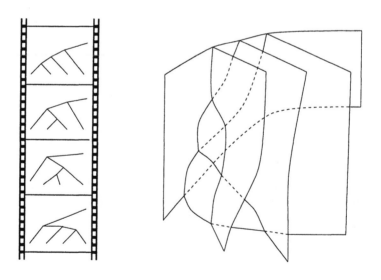

Figure 5: A movie of 6j-symbols and the Elliott-Biedenharn identity (right hand side)

dual to the union of tetrahedra — two tetrahedra glued along a face for Figure 4 and three tetrahedra glued along an edge for Figure 5. These two complexes are related by a deformation that is also one of the Matveev moves [25, 16, 32].

Thus the identities expressed among the $6j$-symbols are diagrammatically expressed as deformations between 2-dimensional cell complexes. So the use of $6j$-symbols in the construction of the Turaev-Viro invariant appears quite natural in the context of the representations and their diagrammatic realizations.

However, there are two obstacles to overcome before the definition of the Turaev-Viro invariant can be made. First, the invariant is a sum over representations, and since we have irreducible representations for *all* non-negative half-integers j, such a sum would be infinite, so we cannot use the classical theory of $U(sl(2))$ to obtain an invariant. Second, the $6j$-symbols that we have defined do not possess full tetrahedral symmetry, and thus they cannot be associated to tetrahedra in any meaningful way. We will overcome the first obstacle in Section 4, by passing to the representations of $U_q(sl(2))$ for q a root of unity. Professor Biedenharn informs us that this is the physicist's notion of renormalization since we are converting an infinite sum to a finite sum. We will overcome the second obstacle by normalizing the $6j$-symbols and by showing that the normalized versions possess the desired symmetry while still satisfying orthogonality and Elliott-Biedenharn relations.

2.7 Computations.   We now express the $6j$-symbols in terms of evaluations of certain spin-nets. In particular, we determine some of the symmetry properties of the $6j$-symbols in the current normalization, and we find a normalization that has full tetrahedral symmetry.

2.7.1 Topological invariance. Let a spin-net be given. This network is a graph with its edges labeled by non-negative half-integers that has only 2-valent and 3-valent vertices. Two edges incident at a 2-valent vertex must have the same label, and if edges with labels a, b, and j are incident at a vertex, then the triple (a, b, j) is admissible. Choose an embedding of the spin-net into a rectangle such that the endpoints of free edges (if any) are on the top and bottom of the rectangle. Suppose that the labels of the edges that appear on the top are $a_1, \ldots, a_n$, and those that appear on the bottom are $b_1, \ldots, b_m$ where these labels are read from left to right once the rectangle has been embedded in the plane.

The embedding of the spin-net allows us to define a map

$$\Box : V^{b_1} \otimes \cdots \otimes V^{b_m} \to V^{a_1} \otimes \cdots \otimes V^{a_n}$$

between tensor products of irreducible representations where projectors $+$ are associated to the 2-valent vertices and maps Y and $\mathsf{\Lambda}$ are associated to the 3-valent vertices. At the bottom of the rectangle the V^bs are mapped via ϕ into the tensor powers of the fundamental representation, and at the top the tensor powers are projected back onto the V^as. Some care has to be taken with regard to the indices along the edges as we indicated in the definition given in Section 2.6.3. Finally, the embedding of the graph is to be in general position with respect to the height function on the rectangle.

2.7.2 Lemma. *The map $\Box$ does not depend on the isotopy class (rel. boundary) of the embedding of the given graph in the rectangle.*

Proof. This follows from the remarks 2.6.5. $\square$

2.7.3 Lemma. *The value of $\overset{n}{\cap}: (V^{1/2})^{\otimes n} \otimes (V^{1/2})^{\otimes n} \to \mathbb{C}$ is given by*

$$\overset{n}{\cap}\left((x_1 \otimes x_2 \otimes \cdots \otimes x_n) \otimes (\bar{x}_n \otimes \cdots \otimes \bar{x}_2 \otimes \bar{x}_1)\right) =$$

$$= \begin{cases} 0 & \text{if } x_j = \bar{x}_j \text{ for some } j = 1, \ldots n, \\ i^n(-1)^{\#\{k:x_k=y\}} & \text{if } \{x_j, \bar{x}_j\} = \{x, y\} \text{ for all } j = 1, \ldots, n \end{cases}$$

where $x_j, \bar{x}_j \in \{x, y\}$ for all $j = 1, \ldots, n$.

Proof. This follows by induction, let us exemplify the formula. In case $n = 2$, the non-zero values of $\overset{2}{\cap}$ are as follows:

$$\overset{2}{\cap}(x \otimes x \otimes y \otimes y) = -1$$

$$\overset{2}{\cap}(x \otimes y \otimes x \otimes y) = 1$$

$$\overset{2}{\cap}(y \otimes x \otimes y \otimes x) = 1$$

$$\overset{2}{\cap}(y \otimes y \otimes x \otimes x) = -1.$$

$\square$

2.7.4 Lemma. *For $u + v = j$, consider the map $\mu_j \circ \overset{j}{\underset{ab}{\bigwedge}} \circ$ $\phi_a \otimes \phi_b : V^a \otimes V^b \to V^j$. We have*

$$\mu_j\left(\overset{j}{\underset{ab}{\bigwedge}} (\phi_a \otimes \phi_b (e_{a,u} \otimes e_{b,v}))\right)$$

$$= i^{a+b-j}(-1)^{a-u}\frac{(a+b-j)!(a+u)!(b+v)!}{(2a)!(2b)!}e_{j,j}$$

Proof. Recall that $e_{a,u} = x^{a+u}y^{a-u}$, that a similar formula for $e_{b,v}$ holds, and that the image under ϕ of a weight vector, $e_{j,t}$, is a symmetrized version of it.

We have to count the number of terms in the sum for $\phi_a(e_{a,u}) \otimes$ $\phi_b(e_{b,v})$ that contribute to the network evaluation. A term that contributes is determined by a subset $S \subset \{1, 2, \ldots, a+b-j\}$ such that $|S| = b - v$ and is of the form

$$\frac{1}{(2a)!}[\underbrace{x \otimes \cdots \otimes x}_{a+j-b} \otimes \bar{x}_1^S \otimes \cdots \otimes \bar{x}_{a+b-j}^S]$$

$$\otimes \frac{1}{(2b)!}[x_{a+b-j}^S \otimes \cdots \otimes x_1^S \otimes \underbrace{x \otimes \cdots \otimes x}_{b+j-a}]$$

where

$$\bar{x}_\ell^S = \begin{cases} x & \text{if } \ell \in S \\ y & \text{if } \ell \notin S \end{cases}$$

and

$$x_\ell^S = \begin{cases} y & \text{if } \ell \in S \\ x & \text{if } \ell \notin S \end{cases}$$

There are $\binom{a+b-j}{b-v}$ such subsets S, and each such subset, S, will arise $(a+u)!(a-u)!(b+v)!(b-v)!$ times coming from $(a+u)!(a-u)!$ of the permutations of $\{1, \ldots 2a\}$ and from $(b+v)!(b-v)!$ of the permutations of $\{1, \ldots 2b\}$. Now

$$\mu_j\left(\left. \vphantom{\Big|} \right|_{a+j-b} \otimes \overset{a+b-j}{\cap} \otimes \left.\vphantom{\Big|}\right|_{b+j-a} (w) \right) = i^{a+b-j}\frac{(-1)^{a-u}}{(2a)!(2b)!}x^{2j}.$$

Therefore,

$$\mu_j\left(\overset{j}{\underset{ab}{\wedge}} (\phi_a \otimes \phi_b (e_{a,u} \otimes e_{b,v})) \right) =$$

$$= i^{a+b-j}(-1)^{a-u}\frac{(a+u)!(a-u)!(b+v)!(b-v)!}{(2a)!(2b)!}\binom{a+b-j}{b-v}x^{2j}.$$

as required. $\square$

2.7.5 Lemma. *Let*

$$\chi_{ba}^{ab} : (V^{1/2})^{\otimes 2b} \otimes (V^{1/2})^{\otimes 2a} \to (V^{1/2})^{\otimes 2a} \otimes (V^{1/2})^{\otimes 2b}$$

denote the map that switches factors:

$$\chi_{ba}^{ab}(x_1 \otimes \cdots \otimes x_{2b}) \otimes (y_1 \otimes \cdots \otimes y_{2a}) = (y_1 \otimes \cdots \otimes y_{2a}) \otimes (x_1 \otimes \cdots \otimes x_{2b}).$$

1.
$$\chi_{ba}^{ab} \circ \overset{ba}{\underset{j}{\mathsf{Y}}} = (-1)^{a+b-j}\, \overset{ab}{\underset{j}{\mathsf{Y}}}.$$

$$\overset{2a \quad 2b}{\underset{2j}{\diamondsuit}} = (-1)^{a+b-j}\, \overset{2a \quad 2b}{\underset{2j}{\mathsf{Y}}}$$

2.
$$\overset{j}{\underset{ba}{\mathsf{\Lambda}}} \circ \chi_{ab}^{ba} = (-1)^{j-a-b}\, \overset{j}{\underset{ab}{\mathsf{\Lambda}}}.$$

$$\overset{2j}{\underset{2a \quad 2b}{\diamondsuit}} = (-1)^{j-a-b}\, \overset{2j}{\underset{2a \quad 2b}{\mathsf{\Lambda}}}$$

3.

$$\left(|_{2m} \otimes \overset{ab}{\underset{j}{\mathsf{Y}}} \right) \circ \chi_{jm}^{mj}$$

$$= (\chi_{am}^{ma} \otimes |_{2b}) \circ \left(|_{2a} \otimes \chi_{bm}^{mb} \right) \circ \left(\overset{ab}{\underset{j}{\mathsf{Y}}} \otimes |_{2m} \right).$$

4.

$$\chi_{jm}^{mj} \circ \left(\overset{j}{\underset{ab}{\lambda}} \otimes \mid_{2m} \right)$$

$$= \left(\mid_{2m} \otimes \overset{j}{\underset{ab}{\lambda}} \right) \circ \left(\chi_{am}^{ma} \otimes \mid_{2b} \right) \circ \left(\mid_{2a} \otimes \chi_{bm}^{mb} \right).$$

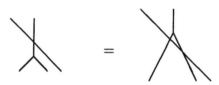

Proof. Part (1) is a direct calculation that can be achieved by evaluating either side on the highest weight vector, $\phi_j(x^{2j})$. Part (2) follows from (1) and Lemma 2.7.2 by rotating the vertex of the right side of the diagram for (2) 180°. Parts (3) and (4) follow by evaluating on the tensor products of appropriate weight vectors. $\square$

2.7.6 Theorem.

$$\mu_k \left(\overset{k}{\underset{ab}{\lambda}} \left(\overset{ab}{\underset{j}{Y}} \left(\phi_j \left(x^{2j} \right) \right) \right) \right) = \left[\Theta(a,b,j) \delta_j^k / \Delta_j \right] (x^{2j})$$

where δ_j^k is a Kronecker δ function, $\Delta_j = (-1)^{2j}(2j+1)$, and

$$\Theta(a,b,k) =$$

$$(-1)^{a+b+k} \frac{(a+b-k)!(a-b+k)!(-a+b+k)!(a+b+k+1)!}{(2a)!(2b)!(2k)!}.$$

Proof. Since the given composition maps an irreducible representation V^j into V^k, it is 0 when $k \neq j$. And when $k = j$ it is a constant multiple of the identity. In Theorem 2.5.5 we computed the value of $\curlyvee$ on a highest weight vector, and in Lemma 2.7.4 we computed $\curlywedge$. Combining these results we obtain

$$\mu_j \left({\curlywedge}_{ab}^{\,j} \left({\curlyvee}_{\,j}^{\,ab} \left(\phi_j \left(x^{2j} \right) \right) \right) \right)$$

$$= \mu_j \left({\curlywedge}_{ab}^{\,j} \left(i^{a+b-j} \sum_{u+v=j} (-1)^{a-u} \binom{a+b-j}{a-u} \cdot \right. \right.$$

$$\phi_a(x^{a+u} y^{a-u}) \otimes \phi_b(x^{b+v} \otimes y^{b-v}) \Big) \Big)$$

$$= x^{2j} \sum_{u+v=j} i^{a+b-j} i^{a+b-j} (-1)^{a-u} \binom{a+b-j}{a-u} \cdot$$

$$(-1)^{a-u} \frac{(a+b-j)!(a+u)!(b+v)!}{(2a)!(2b)!}$$

$$= x^{2j} (-1)^{a+b-j} \frac{((a+b-j)!)^2}{(2a)!(2b)!} \sum_{u+v=j} \frac{(a+u)!(b+v)!}{(a-u)!(b-v)!}$$

$$= x^{2j} (-1)^{a+b-j} \frac{((a+b-j)!)^2}{(2a)!(2b)!} \frac{(a+b+j+1)!(a+j-b)!(b+j-a)!}{(2j+1)!(a+b-j)!}$$

$$= x^{2j} (-1)^{a+b-j} \frac{(a+b-j)!(a+j-b)!(b+j-a)!(b+j+a+1)!}{(2a)!(2b)!(2j+1)!}$$

The next to the last equality is a combinatoric identity. The proof that follows was indicated to us by Rhodes Peele. Consider the set, B, that consists of bijections $f : \{1, 2, \ldots, a+b+j+1\} \to \{1, 2, \ldots, a+b+j+1\}$ such that the value $f(a+j-b+1)$ is greater than every element of $f(\{1, \ldots, a+j-b\})$ and less than every element of $f(\{a+j-b+2, \ldots, 2j+1\})$. We count the elements of the set B in two ways.

There are $\binom{a+b+j+1}{2j+1}$ possible choices for the image $f(\{1,\ldots,2j+1\})$, and each such choice can be arranged in $(a+j-b)!(b+j-a)!$ distinct ways while $f(\{2j+2,\ldots,a+b+j+1\})$ can be arranged in $(a+b-j)!$ distinct ways. Thus

$$\#B = \binom{a+b+j+1}{2j+1}(a+b-j)!(b+j-a)!(a+j-b)!$$

Alternatively, $B = \cup_\ell B_\ell$ where $B_\ell = \{f \in B : f(a+j-b+1) = \ell+1\}$, so $\#B = \sum_\ell \#B_\ell$. There are $\binom{a+b-j}{\ell-(a+j-b)}$ possibilities for the set $f^{-1}(\{1,\ldots,\ell\})$ since the inverse image must exclude the integers in the closed interval $[a+j-b+1, 2j+1]$ and include the integers in the closed interval $[1, a+j-b]$. Each such set can be arranged in $\ell!$ distinct ways. Furthermore, $f^{-1}(\{\ell+2,\ldots,a+b+j+1\})$ is determined by $f^{-1}(\{1,\ldots,\ell\})$ and this can be arranged in $(a+b+j-\ell)!$ distinct ways. So that

$$\#B_\ell = \binom{a+b-j}{\ell-a-j+b}\ell!(a+b+j-\ell)!,$$

and

$$\#B = \sum_\ell \binom{a+b-j}{\ell-a-j+b}\ell!(a+b+j-\ell)!.$$

Now let $\ell = a+u$, and let $j = u+v$, we have

$$\#B = \sum_{u+v=j} \binom{a+b-j}{b-v}(a+u)!(b+v)!$$

$$= \binom{a+b+j+1}{2j+1}(a+j-b)!(b+j-a)!(a+b-j)!.$$

The required identity follows by rearranging the factors in the above equation. This completes the proof. $\square$

The spin-net version of Theorem 2.7.6 is indicated in Corollary 2.7.7. Here we have labeled the spin-nets with the number of strings (= number of tensor factors of $V^{1/2}$ involved). The loop closure of +_{2j} gives the value $\Delta_{2j} = (-1)^{2j}(2j+1)$ which is the denominator of the right hand side. The value Δ_{2j} gives the special case when $a = 0$ of Theorem 2.7.6. The closed network in the numerator on the right has the value $\Theta(a,b,j)$ — this is why the function is named theta.

2.7.7 Corollary. *Let (a,b,j) denote an admissible triple of half-integers. Then we have equality between the following spin-nets.*

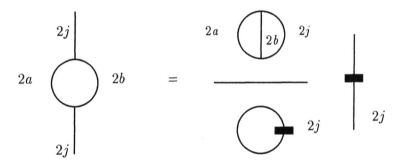

2.7.8 Lemma. *The $6j$ symbols possess the following symmetry*

$$\begin{Bmatrix} m & p & u \\ t & s & r \end{Bmatrix} = \begin{Bmatrix} t & s & u \\ m & p & r \end{Bmatrix}$$

Proof. Embed the spin-net $\overset{m}{\underset{s}{\diagdown}}\!\!r\!\!\overset{p}{\underset{t}{\diagup}}$ in a rectangle with the edges labeled m and p attached to the top edge while the edges labeled s and t are attached to the bottom. By Lemma 2.7.2, the evaluation of the spin-net remains the same when the cross bar of the $\overset{m}{\underset{s}{\diagdown}}\,\overset{}{\underset{}{r}}\,\overset{p}{\underset{t}{\diagup}}$ is rotated clockwise 180° while the end points of the boundary remain fixed.

Having performed such a rotation, recouple; the 6j-symbol that appears is $\left\{ \begin{array}{ccc} t & s & u \\ m & p & r \end{array} \right\}$, where a sum is being taken over u. Then rotate the (now vertical) cross bar 180° counterclockwise. We obtain,

$$\sum_u \left\{ \begin{array}{ccc} m & p & u \\ t & s & r \end{array} \right\} \; \overset{m\diagdown\;\diagup p}{\underset{s\diagup\;\diagdown t}{\mid u}} \; = \sum_u \left\{ \begin{array}{ccc} t & s & u \\ m & p & r \end{array} \right\} \; \overset{m\diagdown\;\diagup p}{\underset{s\diagup\;\diagdown t}{\mid u}} \;.$$

Thus the coefficients are equal since $\mu_m \otimes \mu_p \circ \; \overset{m\diagdown\;\diagup p}{\underset{s\diagup\;\diagdown t}{\mid u}} \; \circ \phi_s \otimes \phi_t$ forms a basis for the set of $U(sl(2))$ invariant maps $V^s \otimes V^t \to V^m \otimes V^p$.
□

2.7.9 Lemma.

$$\frac{\Theta(s,t,k)}{\Delta_k} \left\{ \begin{array}{ccc} m & p & k \\ t & s & r \end{array} \right\} \;=\; \frac{\Theta(m,r,s)}{\Delta_m} \left\{ \begin{array}{ccc} p & k & m \\ s & r & t \end{array} \right\} \quad (1)$$

$$=\; \frac{\Theta(r,t,p)}{\Delta_p} \left\{ \begin{array}{ccc} k & m & p \\ r & t & s \end{array} \right\} \quad (2)$$

Proof. Any of these constants is the coefficient Z in the equation

$$\left(\overset{}{\underset{2m}{+}} \otimes \overset{}{\underset{2p}{+}} \right) \circ \overset{m}{\underset{s}{\diagdown}}\,\overset{}{\underset{}{r}}\,\overset{p}{\underset{t}{\diagup}} \circ \overset{s\;t}{\underset{k}{\mathsf{Y}}} = Z \; \overset{m\;p}{\underset{k}{\mathsf{Y}}}.$$

That there is such a constant Z follows since the space of $U(sl(2))$ invariant maps $V^k \to V^m \otimes V^p$ is 1-dimensional and is spanned by $\mu_m \otimes \mu_p \circ \overset{mp}{\underset{k}{\mathsf{Y}}} \circ \phi_k$. We leave it to the reader to draw the corresponding diagrams. □

2.7.10 Lemma. *Let* $\mathrm{TET}(a, b, c, d, e, f)$ *denote the value of the spin-net depicted below:*

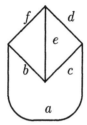

Then,

$$\left\{ \begin{array}{ccc} a & b & f \\ e & d & c \end{array} \right\} = \frac{\mathrm{TET}(a, b, c, d, e, f)\Delta_f}{\Theta(a, b, f)\Theta(d, e, f)}.$$

Proof. The proof follows by recoupling and then applying Corollary 2.7.7 and Lemma 2.7.9 (cf. [16], p. 448). A sketch of the diagrammatic proof is shown below.

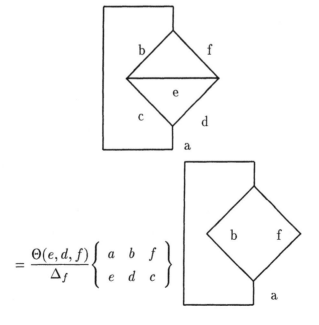

$$= \frac{\Theta(e,d,f)\Theta(a,b,f)}{\Delta_f} \left\{ \begin{array}{ccc} a & b & f \\ e & d & c \end{array} \right\}. \quad \Box$$

2.7.11 Remark. In [18] Kauffman and Lins give a closed formula for the value $\mathrm{TET}(a,b,c,d,e,f)$ in both the case at hand and in the quantum case. Furthermore, there are methods to compute the values of the $6j$-symbols based on the Elliott-Biedenharn identity [2] or the recursive properties of the projectors [24]. In Section 2.8, we define four fundamental $6j$-symbols, and use their values and the Elliott-Biedenharn identity to compute the values of the $6j$-symbols in general.

2.7.12 Lemma. *The symbol,*

$$\left[\begin{array}{ccc} a & b & f \\ e & d & c \end{array} \right] = \frac{(-1)^{a+b+d+e}}{2f+1} \sqrt{\left| \frac{\Theta(a,b,f)\Theta(d,e,f)}{\Theta(a,c,d)\Theta(b,c,e)} \right|} \left\{ \begin{array}{ccc} a & b & f \\ e & d & c \end{array} \right\}$$

$$= \mathrm{TET}(a,b,c,d,e,f)/\sqrt{|\Theta(a,b,f)\Theta(d,e,f)\Theta(a,c,d)\Theta(b,c,e)|}$$

is invariant under all permutations of its columns and under the exchange of any pair of elements in the top row with the corresponding pair in the bottom row. Equivalently, the symbol
$$\left[\begin{array}{ccc} a & b & f \\ e & d & c \end{array} \right]$$
is invariant under the permutations of the set

$$\{\{a,b,f\},\{a,c,d\},\{b,c,e\},\{d,e,f\}\}.$$

Proof. This set is the set of vertices of a tetrahedron with edges labeled by half-integers a,b,c,d,e,f, such that any element in the set forms an admissible triple. We choose an embedding of the 1-skeleton of this tetrahedron into a rectangle (see for example the diagram below). This embedded labeled graph is a spin-net and as such determines a $U(sl(2))$ invariant map $\mathbf{C} \to \mathbf{C}$ (such a map is multiplication by a complex number). There are two embeddings up to isotopy in the sphere; these are determined by the two possible orientations of the tetrahedron. Lemma 2.7.2 gives that isotopic embeddings will result in the same number since passing arcs over infinity in a closed network case does not affect the spin-net evaluation. Lemma 2.7.5 indicates how the spin net evaluation changes when a crossing is introduced. Changing orientations can be achieved by twisting the trivalent vertices, and by pulling strings through as indicated in the diagrams below. Here we have indicated crossing information so that the reader can see how to pull the diagram on the right out to reverse the orientation of the projected tetrahedron. Therefore, the value of the tetrahedral network is independent of the chosen embedding.

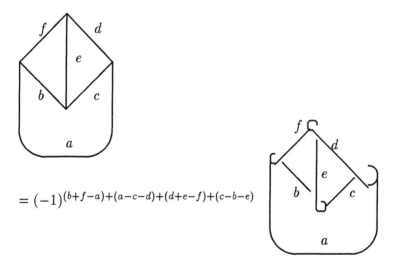

$$= (-1)^{(b+f-a)+(a-c-d)+(d+e-f)+(c-b-e)}$$

By Lemma 2.7.10, the *normalized 6j coefficient* $\begin{bmatrix} a & b & f \\ e & d & c \end{bmatrix}$ is the value of the embedded tetrahedral network divided by the factor $\sqrt{|\Theta(a,b,f)\Theta(d,e,f)\Theta(a,c,d)\Theta(b,c,e)|}$ which possesses tetrahedral symmetry by definition. This completes the proof. $\square$

2.7.13 Theorem. *The orthogonality relation and the Elliott-Biedenharn relation hold for the normalized $6j$-coefficients in the following form.*

Orthogonality,

$$\sum_j (2j+1)(2m+1) \begin{bmatrix} b & c & j \\ k & a & n \end{bmatrix} \begin{bmatrix} a & b & m \\ c & k & j \end{bmatrix} = \delta_{m,n}.$$

Elliott-Biedenharn:

$$\begin{bmatrix} c & d & h \\ g & e & f \end{bmatrix} \cdot \begin{bmatrix} b & h & k \\ g & a & e \end{bmatrix} =$$

$$\sum_j (-1)^z (2j+1) \begin{bmatrix} b & c & j \\ f & a & e \end{bmatrix} \cdot \begin{bmatrix} j & d & k \\ g & a & f \end{bmatrix} \cdot \begin{bmatrix} c & d & h \\ k & b & j \end{bmatrix}$$

where $z = a + b + c + d + e + f + g + h + j + k$

Proof. These results follow from Theorems 2.6.6 and 2.6.7 by substitution, and canceling Θs on either side of the equations. □

2.7.14 Theorem.

1.

$$\begin{Bmatrix} a & c & n \\ b & k & j \end{Bmatrix} =$$

$$\sum_m (-1)^{a+b+c+k-j-n-m} \begin{Bmatrix} a & b & m \\ c & k & j \end{Bmatrix} \begin{Bmatrix} a & c & n \\ k & b & m \end{Bmatrix}.$$

2.

$$\begin{bmatrix} a & c & n \\ b & k & j \end{bmatrix} =$$

$$\sum_m (-1)^{j+m+n} (2m+1) \begin{bmatrix} a & b & m \\ c & k & j \end{bmatrix} \begin{bmatrix} a & c & n \\ k & b & m \end{bmatrix}.$$

3.

$$\sum_n (-1)^{f+j+n+p} \begin{Bmatrix} a & c & n \\ b & k & j \end{Bmatrix} \cdot \begin{Bmatrix} a & f & m \\ d & n & c \end{Bmatrix}$$

$$\cdot \begin{Bmatrix} b & m & p \\ d & k & n \end{Bmatrix}$$

$$= \sum_n (-1)^{k+c+n+m} \begin{Bmatrix} b & f & n \\ d & j & c \end{Bmatrix} \cdot \begin{Bmatrix} a & n & p \\ d & k & j \end{Bmatrix}$$

$$\cdot \begin{Bmatrix} a & f & m \\ b & p & n \end{Bmatrix} .$$

4.

$$\sum_n (-1)^{2n}(2n+1) \begin{bmatrix} a & c & n \\ b & k & j \end{bmatrix} \cdot \begin{bmatrix} a & f & m \\ d & n & c \end{bmatrix}$$

$$\cdot \begin{bmatrix} b & m & p \\ d & k & n \end{bmatrix}$$

$$= \sum_n (-1)^{2n}(2n+1) \begin{bmatrix} b & f & n \\ d & j & c \end{bmatrix} \begin{bmatrix} a & n & p \\ d & k & j \end{bmatrix}$$

$$\cdot \begin{bmatrix} a & f & m \\ b & p & n \end{bmatrix} .$$

5. $\qquad |2a|2b = \sum_j \dfrac{\Delta_j}{\Theta(a,b,j)} \ {}_a^a \bigvee_{\wedge}^b {}_b j \ .$

6.

$$(-1)^{a+b-k}\Theta(a,b,k) = \sum_j (-1)^{a+b-j}\Theta(a,b,j) \begin{Bmatrix} a & b & j \\ a & b & k \end{Bmatrix} .$$

7. $\qquad 1 = \sum_j (-1)^{2j}(2j+1) \begin{bmatrix} a & b & j \\ a & b & k \end{bmatrix}$

Proof for part (1).

$$\cdots = (-1)^{b+c-j} \cdots$$

$$= (-1)^{b+c-j} \sum_n \left\{ \begin{array}{ccc} a & c & n \\ b & k & j \end{array} \right\} \cdots$$

Meanwhile,

$$\cdots = \sum_m \left\{ \begin{array}{ccc} a & b & m \\ c & k & j \end{array} \right\} \cdots$$

$$= \sum_m (-1)^{a+b-m} \left\{ \begin{array}{ccc} a & b & m \\ c & k & j \end{array} \right\} \cdots$$

$$= \sum_m \sum_n (-1)^{a+b-m} \left\{ \begin{array}{ccc} a & b & m \\ c & k & j \end{array} \right\} \left\{ \begin{array}{ccc} a & c & n \\ k & b & m \end{array} \right\} \cdots$$

$$= \sum_n \sum_m (-1)^{a+b-m} \left\{ \begin{matrix} a & b & m \\ c & k & j \end{matrix} \right\} \left\{ \begin{matrix} a & c & n \\ k & b & m \end{matrix} \right\}$$

$$= \sum_{n,m} (-1)^{a+2b-m+n-k} \left\{ \begin{matrix} a & b & m \\ c & k & j \end{matrix} \right\} \left\{ \begin{matrix} a & c & n \\ k & b & m \end{matrix} \right\} \qquad \square$$

Proof for part (3).

$$= (-1)^{b+c-j} \sum_n \left\{ \begin{matrix} a & c & n \\ b & k & j \end{matrix} \right\}$$

$$= \sum_n (-1)^{b+c-j} (-1)^{b+n-k} \left\{ \begin{matrix} a & c & n \\ b & k & j \end{matrix} \right\}$$

$$= \sum_n (-1)^{c+k-j-n} \left\{ \begin{matrix} a & c & n \\ b & k & j \end{matrix} \right\}$$

For convenience, we redraw these diagrams with a different height function:

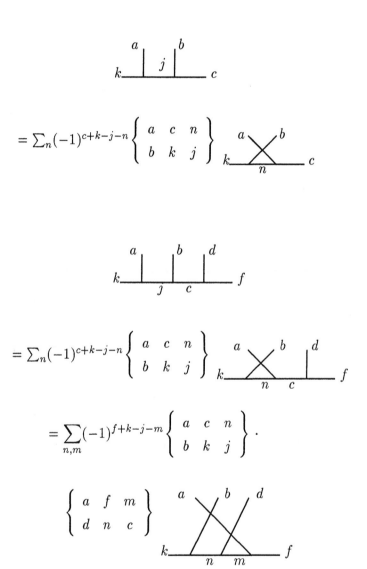

$$= \sum_n (-1)^{c+k-j-n} \left\{ \begin{array}{ccc} a & c & n \\ b & k & j \end{array} \right\}$$

$$= \sum_n (-1)^{c+k-j-n} \left\{ \begin{array}{ccc} a & c & n \\ b & k & j \end{array} \right\}$$

$$= \sum_{n,m} (-1)^{f+k-j-m} \left\{ \begin{array}{ccc} a & c & n \\ b & k & j \end{array} \right\} .$$

$$\left\{ \begin{array}{ccc} a & f & m \\ d & n & c \end{array} \right\}$$

$$= \sum_{n,m,p} (-1)^{f+2k-j-n-p} \begin{Bmatrix} a & c & n \\ b & k & j \end{Bmatrix} \begin{Bmatrix} a & f & m \\ d & n & c \end{Bmatrix}.$$

$$\begin{Bmatrix} b & m & p \\ d & k & n \end{Bmatrix}$$

On the other hand,

$$= \sum_n (-1)^{j+f-c-n} \begin{Bmatrix} b & f & n \\ d & j & c \end{Bmatrix}$$

$$= \sum_{n,p} (-1)^{f+k-c-p} \begin{Bmatrix} b & f & n \\ d & j & c \end{Bmatrix} \begin{Bmatrix} a & n & p \\ d & k & j \end{Bmatrix}$$

$$= \sum_{n,m,p} (-1)^{2f+k-c-n-m} \begin{Bmatrix} b & f & n \\ d & j & c \end{Bmatrix} \begin{Bmatrix} a & n & p \\ d & k & j \end{Bmatrix}.$$

$$\begin{Bmatrix} a & f & m \\ b & p & n \end{Bmatrix}$$

Therefore, comparing the coefficients for fixed m and p, we have:

$$\sum_n (-1)^{f+j+n+p} \begin{Bmatrix} a & c & n \\ b & k & j \end{Bmatrix} \begin{Bmatrix} a & f & m \\ d & n & c \end{Bmatrix} \begin{Bmatrix} b & m & p \\ d & k & n \end{Bmatrix}$$

$$= \sum_n (-1)^{k+c+n+m} \begin{Bmatrix} b & f & n \\ d & j & c \end{Bmatrix} \begin{Bmatrix} a & n & p \\ d & k & j \end{Bmatrix} \begin{Bmatrix} a & f & m \\ b & p & n \end{Bmatrix}.$$

Proof of part (6).

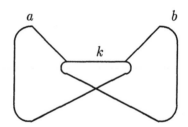

$$= \sum_j \begin{Bmatrix} a & b & j \\ a & b & k \end{Bmatrix}$$

$$= \sum_j (-1)^{a+b-j} \begin{Bmatrix} a & b & j \\ a & b & k \end{Bmatrix}$$

On the other hand,

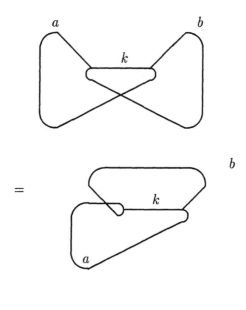

Next we observe that part (2) follows from (1), part (4) follows from (3), and (7) follows from (6) directly. Part (5) follows by recoupling using the $6j$-coefficient $\left\{ \begin{matrix} a & b & j \\ b & a & 0 \end{matrix} \right\}$. This completes the proof. $\square$

2.8 A recursion formula for the $6j$-symbols. In Biedenharn-Louck [2], a recursive method to compute the $6j$-symbols is presented. We summarize this method here.

First, we define the following four $6j$-symbols to be the *fundamental $6j$-symbols*. Their values can be computed by means of the recursion relation for ✚ given in Section 3.5 at $A = 1$. See [24] for details. An alternative computation in the classical case (the case at hand) is given in [18].

$$\left\{ \begin{array}{ccc} x & y & z \\ z+1/2 & 1/2 & x+1/2 \end{array} \right\} = \frac{(x-y+z+1)(x+y+z+2)}{(2x+1)(2z+2)}$$

$$\left\{ \begin{array}{ccc} x & y & z \\ z-1/2 & 1/2 & x+1/2 \end{array} \right\} = \frac{x+y-z+1}{2x+1}$$

$$\left\{ \begin{array}{ccc} x & y & z \\ z+1/2 & 1/2 & x-1/2 \end{array} \right\} = \frac{x-y-z-1}{2z+2}$$

$$\left\{ \begin{array}{ccc} x & y & z \\ z-1/2 & 1/2 & x-1/2 \end{array} \right\} = 1$$

We will compute $\left\{ \begin{array}{ccc} c & d & h \\ g & e & f \end{array} \right\}$ for $e \geq 1/2$ in terms of $\left\{ \begin{array}{ccc} c & d & h \\ \star & e-1/2 & \star \end{array} \right\}$. This pushes everything back to the case $\left\{ \begin{array}{ccc} c & d & h \\ h & 0 & c \end{array} \right\} = 1.$

Recall that the Elliott-Biedenharn identity says

$$\left\{ \begin{array}{ccc} c & d & h \\ g & e & f \end{array} \right\} \cdot \left\{ \begin{array}{ccc} b & h & k \\ g & a & e \end{array} \right\}$$

$$= \sum_j \left\{ \begin{array}{ccc} b & c & j \\ f & a & e \end{array} \right\} \cdot \left\{ \begin{array}{ccc} j & d & k \\ g & a & f \end{array} \right\} \cdot \left\{ \begin{array}{ccc} c & d & h \\ k & b & j \end{array} \right\}.$$

Multiplying by $\left\{ \begin{array}{ccc} a & b & e \\ h & g & k \end{array} \right\}$, summing over k, and using orthogonality, gives

$$\left\{ \begin{array}{ccc} c & d & h \\ g & e & f \end{array} \right\}$$

$$= \sum_{j,k} \left\{ \begin{matrix} b & c & j \\ f & a & e \end{matrix} \right\} \cdot \left\{ \begin{matrix} j & d & k \\ g & a & f \end{matrix} \right\} \cdot \left\{ \begin{matrix} a & b & e \\ h & g & k \end{matrix} \right\} \cdot \left\{ \begin{matrix} c & d & h \\ k & b & j \end{matrix} \right\}.$$

Applying this in the case $a = 1/2$ and $b = e - 1/2$, where admissibility conditions force $j = f \pm 1/2$ and $k = g \pm 1/2$, yields

$$\left\{ \begin{matrix} c & d & h \\ g & e & f \end{matrix} \right\} = A \left\{ \begin{matrix} c & d & h \\ g+1/2 & e-1/2 & f+1/2 \end{matrix} \right\} +$$

$$+ B \left\{ \begin{matrix} c & d & h \\ g+1/2 & e-1/2 & f-1/2 \end{matrix} \right\}$$

$$+ C \left\{ \begin{matrix} c & d & h \\ g-1/2 & e-1/2 & f+1/2 \end{matrix} \right\}$$

$$+ D \left\{ \begin{matrix} c & d & h \\ g-1/2 & e-1/2 & f-1/2 \end{matrix} \right\}$$

where

$$A = \frac{(e+c-f)(g+h-e+1)}{(2e)(2g+1)}$$

$$B = \frac{(e-c+f)(e+c+f+1)(f+d-g)(g+h-e+1)}{(2e)(2f)(2g+1)(2f+1)}$$

$$C = \frac{(e+c-f)(f-d-g)}{(2e)(2g+1)}$$

$$D = \frac{(e-c+f)(e+c+f+1)(f-d+g)(f+d+g+1)}{(2e)(2f)(2f+1)(2g+1)}$$

The coefficients A, B, C, and D are values (with suitable choices for j and k) of the product

$$\left\{ \begin{matrix} e-1/2 & c & j \\ f & 1/2 & e \end{matrix} \right\} \left\{ \begin{matrix} j & d & k \\ g & 1/2 & f \end{matrix} \right\} \left\{ \begin{matrix} 1/2 & e-1/2 & e \\ h & g & k \end{matrix} \right\}$$

which can be evaluated using the four fundamental $6j$-symbols.

2.9 Remarks. In the above discussion, we have reproduced proofs of the important identities among the Clebsch-Gordan coefficients and the $6j$-symbols by means of the spin network analysis of Penrose [27] and Kauffman [16, 18]. The diagrammatic or graphical method seems to be well known [3, 2] but only with the work of [19] and its sequels has this been brought to the fore-front of the theory. The identities 6.16-6.19 given in [19] for the representations of the quantum group are given above (for the classical case) as Theorems 2.6.7, 2.6.6, and parts of Theorem 2.7.14 (but not respectively).

The fact that these proofs are expressible in terms of simple diagrammatic manipulations points to various levels of generalization. In the next section we explore one such level: knot theoretic quantization, in which the braid group replaces the permutation group, and a parameter is inserted into the binor identity. This notion of quantization coincides with the notion of a quantum group as developed by Jimbo [10], Drinfel'd [5], and others (see for example [11] for further references.)

Some puzzles remain about these techniques. Combinatorial identities arise — for example, recursion relations among the Clebsch-Gordan coefficients. Are there simple discrete probabilistic reasons for these formulas, and if so, how are the more elementary combinatorial formulas expressible diagrammatically?

In the next section we turn to the finite dimensional representations of the quantum group $U_q(sl(2))$. We will see the direct analogy with the classical case as the theory develops, and in Section 4, we examine in detail the special case when the quantum parameter is a root of unity.

3 Quantum $sl(2)$

3.1 Some finite dimensional representations. In Section 2 we showed how the tensor products of finite dimensional representations of $sl(2)$ can be decomposed.

In this section we mimic this classical theory in the so-called quantum case where the representation spaces are spaces of homogeneous polynomials in two variables that only commute up to a parameter. It is these representations that give invariants of 3-dimensional manifolds, and physical applications are found in statistical mechanics where they provide solutions to the Yang-Baxter equation.

Recall we have found irreducible representations of $sl(2)$ on $V^j = \{$homogeneous polynomials of degree $2j$ in x and $y\}$. Namely, if $a + b = 2j$

$$E(x^a y^b) = bx^{a+1} y^{b-1}$$

$$F(x^a y^b) = ax^{a-1} y^{b+1}$$

$$H(x^a y^b) = \frac{(a-b)}{2} x^a y^b.$$

Or, with the notation better adapted to the theory of weights,

$$e_{j,m} := x^{j+m} y^{j-m}$$

$$E e_{j,m} = (j - m)e_{j,m+1}$$

$$F e_{j,m} = (j + m)e_{j,m-1}$$

$$H e_{j,m} = m e_{j,m}$$

from which it follows that

$$[E, F]e_{j,m} = 2m e_{j,m}$$

67

We want to "quantize," which, in our context, means introduce a parameter. Roughly speaking, we want to replace integers $n \geq 0$ by

$$[n]_q = q^{n-1} + q^{n-3} + \ldots + q^{-(n-1)} = \frac{q^n - q^{-n}}{q - q^{-1}}.$$

If $q \neq 0$ is fixed, we will write $[n]$ for $[n]_q$. By definition $[0] = 0$ and $[1] = 1$. The theory of the quantum group $U_q(sl(2))$ is particularly interesting in case the quantum parameter q is a root of unity because $[n] = 0$ when $q^{2n} = 1$. This case will be covered in detail in Section 4.

3.1.1 Definition. Let $q \neq 0, 1, -1 \in \mathbf{C}$. *A representation of* $U_q(sl(2))$ *is a vector space W with operators E, F, K (K" $=$ "q^H)* *such that*

$$[E, F] = \frac{K^2 - K^{-2}}{q - q^{-1}} \text{``} = \text{''} \frac{q^{2H} - q^{-2H}}{q - q^{-1}}$$

$$KE = qEK$$

$$KF = q^{-1}FK$$

where $[E, F] = EF - FE$. The *quantum group* $U_q(sl(2))$ is the algebra over $\mathbf{C}$ that is generated by E, F, K, and K^{-1}; the elements are subject to the relations specified above. For the time being, we will not deal with this algebra directly, but instead we will work with it via its representations. Thus we will only consider the entire algebra (and co-algebra) *ex post facto*.

3.1.2 Theorem. *The representations V^j of $sl(2)$ can be quantized in the following sense: Let $q \neq 0, 1, -1 \in \mathbf{C}$. Let $j \in \{0, 1/2, 1, 3/2, \ldots\}$, and let $A \in \mathbf{C}$, where $A^2 = q$. There is a $(2j + 1)$-dimensional representation V_A^j of $U_q(sl(2))$ given abstractly as follows: First, a basis for V_A^j is*

$$\{e_{j,m} : m = j, j - 1, j - 2, \ldots, -j\}.$$

Second, the action of E, F and K is given by

$$Ee_{j,m} = [j-m]e_{j,m+1}$$

$$Fe_{j,m} = [j+m]e_{j,m-1}$$

$$Ke_{j,m} = A^{2m}e_{j,m}$$

from which it follows that

$$[E,F]e_{j,m} = [2m]e_{j,m}.$$

The representation V_A^j is irreducible provided $A^{4r} \neq 1$ for $1 \leq r \leq 2j$.

Proof. Compute directly the actions of EF, FE, KE, EK, KF, and FK. The identities for the Lie brackets follow by manipulating the rational functions of q that result. To prove irreducibility, note that by the hypothesis on r, we have $[r] \neq 0$ for $1 \leq r \leq 2j$. Therefore, the image of any non-zero vector under powers of E and F spans V_A^j. $\square$

One of our present goals (Theorem 3.4.1) is to give concrete realizations of the abstractly given representations V_A^j. The realizations will be generated from the fundamental representations $V_A^{1/2}$ by tensor products. We can take tensor products of quantized representations because the algebra $U_q(sl(2))$ has a comultiplication. Rather than specifying the comultiplication, we will specify the action of $U_q(sl(2))$ on tensor products of representations; from these formulas the comultiplication can be derived.

3.1.3 Theorem. *Let U and V be representations of $U_q(sl(2))$. (As always, $q \neq 0, 1, -1$). Then there is a representation of $U_q(sl(2))$ on $U \otimes V$ given by the following formulas.*

$$E(u \otimes v) = Eu \otimes Kv + K^{-1}u \otimes Ev$$

$$F(u \otimes v) = Fu \otimes Kv + K^{-1}u \otimes Fv$$

$$K(u \otimes v) = Ku \otimes Kv$$

Proof. Three relations must be verified. $\square$

3.1.4 Remarks. The representations for the quantum case will be made explicitly analogous to the those of the classical case. In particular, we will find the quantum analogues of the spaces of homogeneous polynomials as irreducible representations and the identification of these with subrepresentations in the tensor product of copies of the fundamental representation for generic values of the parameter q. These subrepresentations are the images of projectors that are deformations of the projectors in the classical case.

Via such identifications, we will be able to perform the diagrammatic computations that are analogous to those in the classical case. Specifically, we will show in the quantum case how to decompose tensor products of representations (Clebsch-Gordan Theory), and develop the 6j-symbols and verify their important properties. Finally, we will explicitly discuss the case in which A is a root of unity (the case in which the projector may not be defined), and we will show how the diagrammatics allow us to ignore the so-called trace 0 representations that arise.

3.2 Representations of the braid groups. To construct the concrete realizations of the quantum representation spaces, we will find a natural quotient representation of $(V_A^{1/2})^{\otimes 2j}$ that is $U_q(sl(2))$-isomorphic to V_A^j. To describe the quotient, we introduce an action of the braid group $B(2j)$ on $(V_A^{1/2})^{\otimes 2j}$ that commutes with the action of $U_q(sl(2))$. Furthermore, for generic values of q or for q an rth root of unity (where $r > 2j$), we will show that the quotient is isomorphic to a submodule of the tensor power.

Classically, the obvious representation of the symmetric group Σ_{2j} (permuting factors) on $(V^{1/2})^{\otimes 2j}$ commutes with the action of $sl(2)$ on the same space. The subrepresentation on which Σ_{2j} acts trivially, namely the space of symmetric tensors, is isomorphic to the representation V^j of $sl(2)$. Since representations of the braid group are not necessarily semi-simple, we will be forced to consider quotient representations in the quantized situation, rather than subrepresentations.

3.2.1 Definition. The *Artin braid group* $B(n)$ is given by generators and relations as follows:

$$\langle s_1, s_2, \ldots s_{n-1} : \quad s_k s_j = s_j s_k \quad \text{if} \quad |k - j| > 1;$$

$$s_k s_{k+1} s_k = s_{k+1} s_k s_{k+1} \quad \text{if} \quad k = 1, 2, \ldots, n - 2\rangle$$

Remark. The braid group is depicted graphically as indicated below. Multiplication is achieved by vertical juxtaposition of braid diagrams.

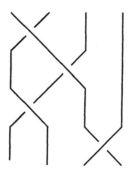

The braid diagram for $s_1 s_2 s_1 s_3^{-1}$

3.2.2 Definitions. Let $A \neq 0 \in \mathbf{C}$ be fixed, and let

$$V_A^{1/2} := \operatorname{span}\{x = e_{1/2,1/2}, y = e_{1/2,-1/2}\}.$$

Here we redefine the matrix representation of the maps $\cup$ and $\cap$ to apply to the quantum case.

Define $\cap_A = \cap : V_A^{1/2} \otimes V_A^{1/2} \to \mathbf{C} = V_A^0$ via

$$x \otimes x \mapsto 0,$$

$$x \otimes y \mapsto iA,$$

$$y \otimes x \mapsto -iA^{-1},$$

and

$$y \otimes y \mapsto 0.$$

Define $\cup_A = \cup : V_A^0 = \mathbf{C} \to V_A^{1/2} \otimes V_A^{1/2}$ by the formula

$$\cup(1) = iAx \otimes y - iA^{-1}y \otimes x.$$

3.2.3 Lemma. *Let $A^2 = q$.*

1. *The maps $\cup_A$ and $\cap_A$ are $U_q(sl(2))$ invariant.*

2.
$$\cap_A \circ \cup_A = \bigcirc_A : \mathbf{C} \to \mathbf{C}$$

is multiplication by $-A^2 - A^{-2} = -[2]$.

3.
$$\cup_A \circ \cap_A = \overset{\cup}{\cap} : V_A^{1/2} \otimes V_A^{1/2} \to V_A^{1/2} \otimes V_A^{1/2}$$

is given by

$$\overset{\cup}{\cap} (x \otimes x) = \overset{\cup}{\cap} (y \otimes y) = 0,$$

$$\overset{\cup}{\cap} (x \otimes y) = -qx \otimes y + y \otimes x,$$

and

$$\overset{\cup}{\cap} (y \otimes x) = x \otimes y - q^{-1}y \otimes x.$$

Proof. This follows by computation. □

3.2.4 Definition. We define a $U_q(sl(2))$ invariant map (called a *positive crossing*)

$$\times_A : V_A^{1/2} \otimes V_A^{1/2} \to V_A^{1/2} \otimes V_A^{1/2}$$

as follows:

$$\times_A = A \left[\; \stackrel{\cup}{\cap} \; \right] + A^{-1} \left[\; | \; \otimes \; | \; \right].$$

Observe that $\left(\times_A \right)^{-1} = A^{-1} \left[\; \stackrel{\cup}{\cap} \; \right] + A \left[\; | \; \otimes \; | \; \right].$

Let the *negative crossing* be defined as follows:

$$\times_A = \left(\times_A \right)^{-1}.$$

As in section 2, the symbol $|$ denotes the identity map, but here the domain is $V_A^{1/2}$.

3.2.5 Theorem. *There is a representation r_A of $B(n)$ on $(V_A^{1/2})^{\otimes n}$ defined by*

$$r_A(s_k) = \underbrace{| \cdots |}_{k-1} \times \underbrace{| \cdots |}_{n-k-1}$$

for $k = 1, 2, \ldots, n-1$.

Proof. It is clear that the images of distant braid generators commute. That the relations $s_k s_{k+1} s_k = s_{k+1} s_k s_{k+1}$ hold in the representation for $k = 1, \ldots, n-2$ is a direct computation which

is usually performed diagrammatically and depends on the fact
that

$$(\cap \otimes \mid) \circ (\mid \otimes \circ \mathrm{X}(\pm)) = (\mid \otimes \cap) \circ (\mathrm{X}(\mp) \otimes \mid)$$

where $\mathrm{X}(\pm)$ denotes the positive or negative crossing depicted
above. (See [16] for the diagrammatic version, or compare with
Lemma 2.3.2.) $\square$

3.2.6 Theorem. *Let $A \in \mathbf{C}$, $q = A^2 \neq 0, 1, -1$. The actions
of $U_q(sl(2))$ and of $B(n)$ (via r_A) on $(V_A^{1/2})^{\otimes n}$ are commuting
actions.*

Proof. This follows because r_A is defined in terms of the $U_q(sl(2))$
maps $\overset{\cup}{\cap}$ and $\mid \otimes \mid$. $\square$

3.3 A finite dimensional quotient of $\mathbf{C}[B(n)]$.   The representations r_A were discovered via the intermediary of certain
finite dimensional algebras that are quotients of the (infinite dimensional) group algebra of $B(n)$.

Next we consider the Temperley-Lieb algebra $TL_n(\delta)$ where
$n \geq 2$ and where $\delta = -(A^2 + A^{-2})$ for some $A \neq 0 \in \mathbf{C}$. Recall
that TL_n is generated by elements I and h_k for $k = 1, \ldots, n-1$
that are subject to the relations given in Section 2.4.

3.3.1 Theorem. *For $n \geq 2$ and $A \neq 0 \in \mathbf{C}$ there is a surjective
algebra homomorphism*

$$\rho_A : \mathbf{C}[B(n)] \to TL_n(-[2])$$

that is defined on generators as follows:

$$\rho_A(s_k) = A^{-1}I + Ah_k$$
$$\rho_A(s_k^{-1}) = AI + A^{-1}h_k$$

Proof. The situation is analogous to Lemma 2.4.2 and Theorem 3.2.5. So details are omitted. □

3.3.2 Remark. The relation

$$h_k^2 = (-A^2 - A^{-2})h_k$$

in $TL_n(-[2])$ cuts the braid algebra down to a finite dimensional algebra.

3.3.3 Remark. The formula for $\rho_A(s_k)$ is referred to as the *bracket identity*. This identity encapsulates Kauffman's simplification of the Jones polynomial [16]. Recall that the bracket identity is given diagrammatically as:

$$\Big\rangle\!\!\!\Big\langle \; = A \Big\langle\; \cup \atop \cap \;\Big\rangle + A^{-1} \Big\langle\; | \; | \;\Big\rangle .$$

Given a knot diagram, $\mathcal{K}$, of a knot, K, the bracket identity is used to compute a polynomial $\langle \mathcal{K} \rangle$ in $A^{\pm 1}$ by removing each crossing via the bracket identity and associating the loop value $-A^{-2} - A^2$ to each of the simple closed curves that results in any of the daughter diagrams. The bracket is an invariant of the regular isotopy class of the diagram, but not of the knot type K. To obtain an invariant of K, define

$$L(K)(A) = (-A)^{-3w(\mathcal{K})}\langle \mathcal{K} \rangle$$

where $w(\mathcal{K})$ is the writhe of the diagram and is computed by orienting the diagram and computing the sum of the signs of the crossings. The Jones polynomial is obtained by setting $A = t^{-1/4}$. See [16] for details.

It is clear that the representation r_A of the braid group defined in Theorem 3.2.5 factors through ρ_A. The algebra $TL_n(-[2])$ is represented on $(V_A^{1/2})^{\otimes n}$ via the map

$$\theta_A : h_k \mapsto \underbrace{|\cdots|}_{k-1} \; \overset{\cup}{\cap} \; \underbrace{|\cdots|}_{n-k-1}$$

as in the classical case but with $\overset{\cup}{\cap} = \overset{\cup}{\underset{A}{\cap}}$. We have the following result:

3.3.4 Theorem. *The representation*

$$\theta_A : TL_n(-[2]) \to \operatorname{Aut}(V_A^{1/2})^{\otimes n}$$

is faithful.

The case when $A = 1$ is covered in Theorem 2.4.3. The case $A = -1$ is the same as $A = 1$ because both $TL_n(-[2])$ and its representation θ_A depend on A only through $q = A^2$. The proof of the generic case and of the case that A is a $4r$th root of unity and $r \geq n$ is explained in Section 4.2. The remaining case is proven in [7]. □

3.3.5 Notation. Let $I_A(n)$ denote the two-sided ideal in $TL_n(-[2])$ that is generated by the set $\{h_1, \ldots, h_{n-1}\}$.

3.3.6 Lemma. *The ideal $I_A(n)$ is a proper ideal, of $\mathbf{C}$-vector space codimension 1 in $TL_n(-[2])$.*

Proof. The space $TL_n(-[2]) = TL_n$ is spanned by 1 and the monomials in the h_k, so the codimension of $I_A(n)$ can only be 0 or 1. We need only prove that $1 \notin I_A(n)$. This is easily seen from the representation of $TL_n(-[2])$ on $(V_A^{1/2})^{\otimes n}$. For $1(x \otimes x \otimes \cdots \otimes x) = x \otimes x \otimes \cdots \otimes x$, but $x \otimes x \otimes \cdots \otimes x \notin I_A(n)(x \otimes x \otimes \cdots \otimes x)$.

3.4 A model for the representations V_A^j. Let

$$\text{Ten}(V_A^{1/2}) = \bigoplus_{j=0,1/2,\ldots} (V_A^{1/2})^{\otimes 2j}$$

be the tensor algebra of $V_A^{1/2}$, and let

$$L_A \subset \text{Ten}(V_A^{1/2})$$

be the two-sided ideal generated by $qx \otimes y - y \otimes x$ (where $q = A^2$, as always), and let

$$L_A^{2j} = L_A \cap (V_A^{1/2})^{\otimes 2j}.$$

Then there is an isomorphism

$$\text{Ten}(V_A^{1/2})/L_A \cong \mathbf{C}[x,y]/(yx - qxy)$$

between the quotient algebra and the algebra of polynomials in the noncommuting variables x and y, where $yx = qxy$. (The parameter q commutes with x and y as it is a member of the ground field.) We will identify these two algebras in the sequel. Moreover, we have

$$\text{Ten}(V_A^{1/2})/L_A = \bigoplus_{j=0,1,\ldots} (V_A^{1/2})^{\otimes 2j}/L_A^{2j}$$

where

$$(V_A^{1/2})^{\otimes 2j}/L_A^{2j}$$

can be identified with the space of homogeneous polynomials of degree $2j$ in x and y, where $yx = qxy$.

A direct computation shows that

$$L_A^{2j} = I_A(2j)(V_A^{1/2})^{\otimes 2j}.$$

Because $I_A(2j)$ is an ideal in $TL_{2j}(-[2])$ whose action on $(V_A^{1/2})^{\otimes 2j}$ commutes with the action of $U_q(sl(2))$, it is clear that L_A^{2j} is a subrepresentation of $(V_A^{1/2})^{\otimes 2j}$ for the actions of all three of TL_{2j}, $B(2j)$, and $U_q(sl(2))$.

3.4.1 Theorem. *The actions of $U_q(sl(2))$, $B(2j)$, and TL_{2j} on $(V_A^{1/2})^{\otimes 2j}/L_A^{2j}$ can be described as follows:*

1. *For TL_{2j}:*

$$Iw = w, \quad h_i w = 0 \qquad \text{for all } w \in (V_A^{2j})^{\otimes 2j}/L_A^{2j}$$

2. *For $B(2j)$:*

$$s_i w = A^{-1} w \qquad \text{for all } w \in (V_A^{1/2})^{\otimes 2j}/L_A^{2j}$$

3. *For $U_q(sl(2))$: There is an isomorphism*

$$\omega_j : V_A^j \to (V_A^{1/2})^{\otimes 2j}/L_A^{2j}$$

   given by
$$\omega_j(e_{j,m}) = A^{(j+m)(j-m)} x^{j+m} y^{j-m}$$

Henceforth, we will identify V_A^j with $(V_A^{1/2})^{\otimes 2j}/L_A^{2j}$ via the isomorphism ω_j.

Proof. Item (1) follows because $h_i((V_A^{1/2})^{\otimes 2j}) \subset L_A^{2j}$. Item (2) follows from (1) and the bracket identity. Item (3) follows from the next Lemma which indicates that in the case at hand E, F, and K act as quantum differential operators; so these actions are analogous to those given in Section 2.2.

3.4.2 Lemma.

1.
$$F(x^a y^b) = [a] A^{a-b-1} x^{a-1} y^{b+1}$$

2.
$$E(x^a y^b) = [b] A^{b-a-1} x^{a+1} y^{b-1}$$

3.
$$K(x^a y^b) = A^{a-b} x^a y^b$$

Proof. Notice that $V_A^{1/2}$ consists of the set of linear combinations of x and y. Thus $Ex = Fy = 0$, while $Fx = y$, $Ey = x$, $Kx = Ax$, and $Ky = A^{-1}x$. These computations form the initial steps of inductions that will follow.

We compute, $K(x^{\otimes a} \otimes y^{\otimes b}) = (Kx)^{\otimes a} \otimes (Ky)^{\otimes b} = A^a x^{\otimes a} \otimes A^{-b} y^{\otimes b} = A^{a-b} x^{\otimes a} \otimes y^{\otimes b}$ where, for example, $x^{\otimes a}$ is the tensor product of a factors of x. Thus the third identity holds.

By Item (1), we mean that $F(x^{\otimes a} \otimes y^{\otimes b}) \equiv [a] A^{a-b-1} x^{\otimes a-1} \otimes y^{\otimes b+1}$ (mod L_A).

First consider the case where $b = 0$. Then induct on a. For $a \geq 2$ we have,

$$F(x^{\otimes a}) = F(x^{\otimes a-1} \otimes x)$$

$$= K^{-1} x^{\otimes a-1} \otimes Fx + F(x^{\otimes a-1}) \otimes Kx$$

$$\equiv [a] A^{a-1} x^{\otimes a-1} \otimes y \quad (\text{mod } L_A)$$

because $A^{1-a} + [a-1] A^{a+1} = [a] A^{a-1}$.

For general $b \geq 1$,

$$F(x^{\otimes a} \otimes y^{\otimes b}) = F(x^{\otimes a} \otimes y^{\otimes b-1} \otimes y)$$

$$= K^{-1}(x^{\otimes a} \otimes y^{\otimes b-1}) \otimes Fy + F(x^{\otimes a} \otimes y^{\otimes b-1}) \otimes Ky$$

$$\equiv [a] A^{a-b} x^{\otimes a-1} \otimes y^b \otimes A^{-1} y \quad (\text{mod } L_A.)$$

The proof of (2) follows along the same lines. This completes the proof of the Lemma $\square$.

3.4.3 Remark. Let χ_A be the one dimensional character of $B(n)$ such that $\chi_A(s_i) = A^{-1}$. (If $A = 1$ or -1 then χ_A factors through the permutation group Σ_n, giving the trivial character $(A = 1)$ or the sign $(A = -1)$. Thus χ_A is a sort of "generalized sign".) Part (2) of Theorem 3.4.1 asserts that the representation of $B(2j)$ on the space of homogeneous polynomials of degree $2j$ in x and y is through the character χ_A.

In the next section, we will define for certain values of A, a $U_q(sl(2))$ invariant projector $+_{2j}^A : (V_A^{1/2})^{\otimes 2j} \to (V_A^{1/2})^{\otimes 2j}$ whose image is a subrepresentation that is isomorphic to V_A^j. The projection is analogous to the projection in the classical case, and from it we will be able to construct the quantum Clebsch-Gordan coefficients and the quantum $6j$-symbols.

3.5 The Jones-Wentzl projectors. In analogy with the classical case, we will define idempotents, called the *Jones-Wentzl projectors*,

$$+_n^A : (V_A^{1/2})^{\otimes n} \to (V_A^{1/2})^{\otimes n}$$

in the algebra of $U_q(sl(2))$ invariant transformations on $(V_A^{1/2})^{\otimes n}$.

3.5.1 Definition. Let $A \neq 0, 1, -1$ denote a complex number, and let $A^2 = q$. There is a canonical embedding of TL_{n-1} into TL_n obtained by juxtaposing a straight string on the right of every generator in TL_{n-1}. We will consider all of the elements in TL_n to also be in TL_{n+k} for the rest of this section. Define *the Jones-Wentzl Projector*, $+_n^A \in TL_n$ via the recursion relation:

$$+_n^A = +_{n-1}^A \otimes | + [n-1]/[n](+_{n-1}^A \otimes |) \circ h_{n-1} \circ (+_{n-1}^A \otimes |)$$

where

$$+_1^A = | = \mathrm{id} : V_A^{1/2} \to V_A^{1/2}.$$

ain $h_n \circ (\dashv_n^A \otimes |) \circ h_n(\dashv_n^A \otimes |) = -[n+1]/[n]h_n(\dashv_n^A \otimes |)$

tity

$$-[2] + \frac{[n-1]}{[n]} = -\frac{[n+1]}{[n]}$$

ntum numbers that is easy to verify. Next we use this

at $\dashv_n^A \dashv_n^A = \dashv_n^A$.

fourth term, the icon that represents $(h_{n-1} \circ (\dashv_{n-1}^A \otimes |))^2$
laced by the icon representing $-[n]/[n-1]h_n(\dashv_{n-1}^A \otimes |_2)$,

Observe that $\dashv_n^A$ is not defined when $[n]! = 0$; thus we assume further that $A^{4r} \neq 1$ for $1 \leq r \leq n$.

3.5.2 Theorem [13]. cf. [18, 23, 16]. *The element* $\dashv_n^A \in TL_n$ *satisfies the conditions:*

1. $h_n \circ (\dashv_n^A \otimes |) \circ h_n \circ (\dashv_n^A \otimes |) = (-1)\frac{[n+1]}{[n]} h_n \circ (\dashv_n^A \otimes |).$

2. $\dashv_n^A \circ \dashv_n^A = \dashv_n^A.$

3. $\dashv_n^A \circ u = u \circ \dashv_n^A = 0$ *for all* $u \in I_A(n)$.

Moreover, any non-zero element in TL_n *satisfying these conditions must be equal to* $\dashv_n^A$.

Proof. The recursion relation above defines *the projector,* $\dashv_n^A$. We will mimic the algebraic proof presented in [16], but here we use diagrams. The diagrammatic version of the recursion relation

$$\dashv_n^A = (\dashv_{n-1}^A \otimes |) + \frac{[n-1]}{[n]}(\dashv_{n-1}^A \otimes |) \circ h_{n-1} \circ (\dashv_{n-1}^A \otimes |)$$

is depicted as follows:

We will assume, by induction, that $\dashv_{n-1}^A \dashv_{n-1}^A = \dashv_{n-1}^A$, and that $\dashv_{n-1}^A h_k = h_k \dashv_{n-1}^A = 0$, for $1 \leq k \leq n-2$ the case of $\dashv_1^A$ being trivial. Then we show that

$$h_n \circ (\dashv_n^A \otimes |) \circ h_n(\dashv_n^A \otimes |) = \frac{(-1)[n+1]}{[n]} h_n \circ (\dashv_n^A \otimes |)$$

by applying h_n to the recursion for $\dashv_n^A$ to obtain $h_n \circ (\dashv_n^A \otimes |) =$

$$\frac{n-1}{n} = \quad + [n-1]/[n] \quad \frac{n-1}{n-2} \quad \frac{n-1}{n-1}$$

Then we obtain, $(h_n \circ (\,\text{\Large$+$}^A_n \otimes |\,))^2 =$

$$= \frac{n-1}{n-1} = \quad + [n-1]/[n] \quad \frac{n-1}{n-2} \quad \frac{n-1}{n-1}$$

$$+[n-1]/[n] \quad \frac{n-1}{n-2} \quad n-1$$

$$+ ([n-1]/[n])^2 \qquad =$$

$$= \quad -[2] \quad \frac{n-1}{} \quad + [n-1]/[n] \quad \frac{n-1}{n-2} \quad \frac{}{n-1}$$

We obta
by the ide

$$-[2][n-1]/[n]$$

$$n$$

$$n -$$

among qu
to show th

$$+ ([n-$$

$$+[n-1]/[$$

The closed loops were replaced
A^{-2}, and the wiggly lines were s
most horizontal line in the second i
by the line above it because $\text{\Large$+$}^A_{n-1}\text{\Large$+$}$
the single vertical strings in the sec
string to the left. In this way, the
from $n-2$ to $n-1$. (The labels on tl
icon are omitted for type-setting rea:
a composition of idempotents. So th
following:

$$(-[2] + [n-1]/[$$

$$\left[\quad \text{\Large$+$} \quad + [n-1]/[n] \quad \right.$$

$$+[n-1]/[$$

$$=$$

$$+[n-1]$$

In the
is rep

and the projectors involving $n - 1$ strings are absorbed since these are idempotents by induction. So far we have shown that $+_n^A$ are idempotents. Furthermore, $+_n^A \neq 0$ because the coefficient of $|_n$ in the sum is 1, as can be seen by induction.

By induction and by this recursion relation, $h_k +_n^A = 0 = +_n^A h_k$ if $k = 1, \ldots, n - 2$. Next we compute $h_{n-1} +_n^A =$

The last sum is obtained again by the fact that $(h_{n-1} +_{n-1}^A)^2 = -[n]/[n - 1]h_{n-1} +_{n-1}^A$ and the sum of these two terms is clearly 0. By the top/bottom symmetry of the diagrams, it follows that $+_n^A h_{n-1} = 0$. Thus we have an inductive construction for the projectors.

Next we show that these elements are unique. Any given element in the Temperley-Lieb algebra can be represented as $\alpha|_n + \mathcal{U}$ where α is in the ground ring, $|_n$ denotes the identity, and $\mathcal{U} \in I_A(n)$ so that $\mathcal{U}$ is a linear combination of products of the h_j. Suppose that $g_n^2 = g_n$ is a non-zero element of TL_n such that $g_n I_A(n) = \{0\}$. Then $g_n = \alpha|_n + \mathcal{U} = \alpha^2|_n + 2\alpha\mathcal{U} + \mathcal{U}^2$. In particular, $\alpha^2 - \alpha = 0$ since $|_n \notin I_A(n)$. If $\alpha = 0$, then $0 = g_n \mathcal{U} = g_n^2 = g_n$. Hence $\alpha = 1$. Thus any non-zero idempotent f_n that kills $I_A(n)$

can be written in the form $f_n = |_n + \mathcal{U}'$. Finally, $g_n = g_n(|_n + \mathcal{U}') = (|_n + \mathcal{U})f_n = f_n$. This proves uniqueness $\square$.

3.5.3 Notation.   For every permutation $\sigma \in \Sigma_n$, define a braid $\hat{\sigma} \in B(n)$ as follows. Write σ in any way as the product of a minimal number of *adjacent transpositions* (*i.e.*, transpositions of the form $\sigma_k = (k, k+1)$), and let $T(\sigma)$ denote this minimal number. Then lift the product to $B(n)$ by replacing each of the transpositions σ_k in the product by the corresponding braid generator s_k. The minimality of the product for σ insures that the lift $\hat{\sigma}$ will depend only on the permutation σ and not on the particular product representation chosen.

Diagrammatically, the transposition σ_k is represented by n arcs running down the page in which the kth and $(k+1)$st arcs cross. The corresponding braid generator s_k can be represented by the same arcs, where the kth arc crosses *over* the $k+1$st.

Next represent the braid group into the algebra of transformations on $(V_A^{1/2})^{\otimes n}$ via the bracket identity. Let $[x]$ denote the quantum integer that corresponds to the integer x. Finally, let $[n]! = [n][n-1] \cdots \cdots [1]$.

3.5.4 Proposition [18].   *The Jones-Wentzl projector is also given by the formula:*

$$\boxed{}_n^A = \frac{A^{n(n-1)}}{[n]!} \sum_{\sigma \in \Sigma_n} (A^{-3})^{T(\sigma)} \hat{\sigma}.$$

Proof of Proposition. We must show that the formula above defines an element that kills each of the h_k, $1 \le k \le n-1$, and that the coefficient of $|_n$ is 1 when the sum is expanded in terms of the standard basis for the Temperley-Lieb algebra. The proof we present follows [18].

First we recall the canonical inductive construction of the set of permutations. Let $(k, \ldots, n)$ denote the cyclic permutation of the elements k through n in Σ_n; this cycle $(k, \ldots, n)$ can be written as a product of adjacent transpositions: $(k, \ldots, n) = (k, k+1)(k+1, k+2) \cdots \cdots (n-1, n)$. To a bijection $t : \{1, \ldots, n\} \to \{1, \ldots, n\}$, we associate a pair (k, t_k), where $t(n) = k$, $t = (k, k+1, \ldots, n)t_k$, and $t_k(n) = n$ so t_k can be regarded as a permutation of $\{1, \ldots, n-1\}$. In this way, the minimal number of adjacent transpositions that it takes to write t is equal to the minimal number that it takes to write t_k plus $n - k$.

Consider an ascending path that starts from the lower left and travels upward through the triangle depicted below.

$$(0,1) \quad (1,2) \quad \cdots \quad (n-2, n-1) \quad (n-1, n)$$
$$(0,1) \quad (1,2) \quad \cdots \quad (n-2, n-1)$$
$$\cdots$$
$$(0,1) \quad (1,2)$$
$$(0,1)$$

A permutation can be represented by such a path as product of the switches to the right of the path on the top row, times the product of the switches to the right on the second row, and so forth. For example, the vertical path through points labeled $(0,1), (0,1), \ldots (0,1)$ is the twist

$$(1,2) \cdots (n-1, n)(1,2) \cdots (n-2, n-1) \cdot \cdots \cdot (1,2),$$

and the identity permutation is represented by the diagonal path $(0,1), (1,2), \ldots, (n-1, n)$. In fact, such descriptions of permutations by paths always use the minimal number of adjacent transpositions necessary to write the permutation.

We compute the coefficient of $|_n$ in the sum $\sum_{\sigma \in \Sigma_n} (A^{-3})^{T(\sigma)} \hat{\sigma}$. Each permutation in the sum contributes a term of the form

$A^{-4T(\sigma)}$ to the coefficient of $|_n$ by the bracket identity. We need the computation

$$A^{\pm n(n-1)}[n]! = \sum_{\sigma \in \Sigma_n} A^{\pm 4T(\sigma)}.$$

To prove this, we use long multiplication to multiply out

$$A^{n(n-1)}[n]! =$$

$$(A^{4(n-1)} + A^{4(n-2)} + \ldots 1)(A^{4(n-2)} + A^{4(n-3)} + \ldots + 1)\cdots(A^4 + 1).$$

The terms in each of the factors are arranged along the horizontals in the triangular array that is depicted above. A term in this expansion corresponds to an ascending path. Moreover, the coefficient of A^{4k}, when like terms are combined, is the number of paths that have k points to their right, and this is the number of permutations that minimally use k adjacent transpositions. Thus the coefficient of $|_n$ in our expression for $\overset{+A}{\underset{n}{\,}}$ is 1.

Now for any given k, the minimal product representations of the permutations can be chosen in such a way that the set of all permutations is partitioned into a set of words W that do not end with $\sigma_k = (k, k+1)$, and the set $W\sigma_k$. (For example, the triangular scheme above does this for $k = 1$.) Clearly, these two sets have the same number of elements, and the number of transpositions that it takes to write an element $w\sigma_k$ is one more than the number of transpositions that is takes to write w for $w \in W$.

The computation that $0 = \left(\sum_{\sigma \in \Sigma_n}(A^{-3})^{T(\sigma)}\hat{\sigma}\right) h_k$ follows because $\hat{\sigma}_k h_k = -A^3 h_k$, and so the contribution of word $w \in W$ is canceled by the contribution of $w\sigma_k$. A similar argument shows that $0 = h_k \left(\sum_{\sigma \in \Sigma_n}(A^{-3})^{T(\sigma)}\hat{\sigma}\right)$. This completes the proof. $\square$

3.5.5 Remark. The formula given above for the projector indicates precisely how $\mathbf{+}_n^A$ is analogous to the classical symmetrizing projection $\mathbf{+}_n$. Moreover, the classical projection satisfies the same recursion relation with quantum integers replaced by integers. Such a replacement is an evaluation of the quantum projector at $A = 1$, and for that value of A, positive braiding is indistinguishable from negative braiding.

When the Jones-Wentzl projector $\mathbf{+}_n^A$ is defined (*e.g.* when A is transcendental), it is a nonzero central idempotent such that $\mathbf{+}_n^A \cdot I_A(n) = (0)$. So the Temperley-Lieb algebra has a direct sum decomposition as scalar multiples of the Jones-Wentzl projector and the ideal $I_A(n)$:

$$TL_n(-[2]) = \mathbf{C}\mathbf{+}_n^A \oplus I_A(n).$$

We have the direct sum decomposition

$$(V_A^{1/2})^{\otimes n} = \mathbf{+}_n^A (V_A^{1/2})^{\otimes n} \oplus L_A^n$$

for representations of $B(n)$ and $U_q(sl(2))$ where

$$L_A^n = I_A(n)(V_A^{1/2})^{\otimes n}$$

is the kernel of the projector $\mathbf{+}_n^A$. This leads us to the following:

3.5.6 Definition. A $U_q(sl(2))$-equivariant map

$$\phi_j = \phi_j^A : V_A^j \to (V_A^{1/2})^{\otimes 2j}$$

is defined in terms of the projector as follows:

$$\phi_j(e_{j,m}) = A^{(j+m)(j-m)}\mathbf{+}_{2j}^A(x^{\otimes j+m} \otimes y^{\otimes j-m}).$$

This formula only makes sense when $[2j]! \neq 0$, and thus the value of the quantum parameter is important. When it is defined, the map ϕ_j is a lift of the map ω_j (which was defined in

Theorem 3.4.1); in other words, $p_j \phi_j = \omega_j$ where $p_j : (V^{1/2})^{\otimes 2j} \rightarrow$ $(V^{1/2})^{\otimes 2j} / L_A^{2j}$ is the projection.

The proposition that follows indicates that the image of ϕ_j is the quantum analogue of the symmetric polynomials in x and y.

3.5.7 Proposition. *For $a + b = n$,*

$$+_n^A(x^{\otimes a} \otimes y^{\otimes b}) = \frac{1}{q^{ab} \begin{bmatrix} n \\ b \end{bmatrix}} \sum_{\substack{S \subset \{1, \ldots, n\} \\ |S| = b}} q^{t_n(S)} x_1^S \otimes \cdots \otimes x_n^S$$

where

$$x_k^S = \begin{cases} x & \text{if} \quad k \notin S \\ y & \text{if} \quad k \in S \end{cases}$$

$t_n(S)$ is the minimal number of adjacent transpositions that it takes to move S to the subset $\{a + 1, \ldots, a + b\}$ ($n = a + b$), and

$$\begin{bmatrix} n \\ b \end{bmatrix} = \frac{[n]!}{[n - b]![b]!}$$

is the quantum analogue of a binomial coefficient (See Lemma 3.6.1 for the corresponding quantum recursion.)

We use the following Lemmas.

3.5.8 Lemma.

$$F(u_1 \otimes \cdots \otimes u_n)$$

$$= \sum_{m=1}^n K^{-1} u_1 \otimes \cdots \otimes K^{-1} u_{m-1} \otimes F u_m \otimes K u_{m+1} \otimes \cdots \otimes K u_n$$

Proof. Induct on n. $\square$

3.5.9 Lemma.

$$F\left(\sum_{\substack{S \subset \{1,\ldots,n\} \\ |S| = b}} q^{t_n(S)} x_1^S \otimes \cdots \otimes x_n^S\right) =$$

$$= [b+1]q^b A^{1-n} \sum_{\substack{T \subset \{1,\ldots,n\} \\ |T| = b+1}} q^{t_n(T)} x_1^T \otimes \cdots \otimes x_n^T$$

where

$$x_k^R = \begin{cases} x & \text{if } k \notin R \\ y & \text{if } k \in R \end{cases}$$

for $R = S, T$.

Proof of Proposition 3.5.7. Assuming Lemma 3.5.9 we induct on b. In case $b = 0$ the result follows by using induction on n and the recursion relation for the projector $+_n^A$.

For general values of b we use the maps $\phi_j : V_A^j \to (V_A^{1/2})^{\otimes 2j}$ defined above, where $j = n/2$. On the one hand,

$$\phi_j(Fe_{j,m}) = F\phi_j(e_{j,m})$$
$$= A^{(j+m)(j-m)} F(+_n^A(x^{\otimes j+m} \otimes y^{\otimes j-m}))$$

where $+_n^A$ is evaluated by means of the inductive hypothesis. On the other hand,

$$\phi_j(Fe_{j,m}) = \phi_j([j+m]e_{j,m-1})$$
$$= [j+m]A^{(j+m-1)(j-m+1)}+_n^A(x^{\otimes j+m-1} \otimes y^{\otimes j-m+1}).$$

The proof will follow by comparison of the two sides of the equation once we establish the following:

Proof of Lemma 3.5.9. Recalling that $Fy = 0$, we compute

$$F\left(\sum_{\substack{S \subset \{1,\ldots,n\} \\ |S| = b}} q^{t_n(S)} x_1^S \otimes \cdots \otimes x_n^S \right)$$

$$= \sum_{\substack{S \subset \{1,\ldots,n\} \\ |S| = b}} \sum_{\substack{m = 1 \\ m \notin S}}^{n} q^{t_n(S)} K^{-1} x_1^S \otimes$$

$$\cdots \otimes K^{-1} x_{m-1}^S \otimes y \otimes K x_{m+1}^S \otimes \cdots \otimes K x_n^S$$

$$= \sum_{\substack{T \subset \{1,\ldots,n\} \\ |T| = b+1}} \left(\sum_{m \in T} q^{t_n(T \setminus \{m\})} A^{f(T,m)} \right) x_1^T \otimes \cdots \otimes x_n^T$$

where

$$f(T,m) = \#\{y \in \{x_1^T, \ldots, x_{m-1}^T\}\} - \#\{x \in \{x_1^T, \ldots, x_{m-1}^T\}\}$$

$$- \#\{y \in \{x_{m+1}^T, \ldots, x_n^T\}\} + \#\{x \in \{x_{m+1}^T, \ldots, x_n^T\}\}.$$

(We recall that $K^{\pm 1} x = A^{\pm 1} x$, and $K^{\pm 1} y = A^{\mp 1} y$.)

Next we show that for $T \subset \{1, \ldots, n\}$,

$$\sum_{m \in T} q^{t_n(T \setminus \{m\})} A^{f(T,m)} = [|T|] q^{|T|-1} A^{1-n} q^{t_n(T)}.$$

from which the proof of Lemma 3.5.9 will follow. To this end, we write $T = T_1 \cup \{m\} \cup T_2$ where $T_1 < m < T_2$. Then

$$f(T,m) = |T_1| - (m - 1 - |T_1|) - |T_2| + (n - m - |T_2|),$$

and

$$t_n(T \setminus \{m\}) = t_n(T) + |T_1| - (n - m - |T_2|).$$

So

$$2t_n(T \setminus \{m\}) + f(T, m) = 2t_n(T) + 4|T_1| - n + 1.$$

Hence

$$\sum_{m \in T} q^{t_n(T \setminus \{m\})} A^{f(T,m)} = A^{2t_n(T)-n+1} \left(\sum_{k=0}^{|T|-1} A^{4k} \right)$$

$$= [|T|] q^{|T|-1} A^{1-n} q^{t_n(T)}$$

as desired. This completes the proof of Lemma 3.5.9. $\square$

3.6 The quantum Clebsch-Gordan theory. Recall that in the classical case of $U(sl(2))$, the Clebsch-Gordan maps are constructed as compositions of $\overset{n}{\cap}$, $\overset{n}{\cup}$, and $+$. Moreover, the computation of these maps relied on some standard combinatorial identities. The same is true in the quantum case. We first establish a recursion relation for the quantum binomial coefficients, and prove some elementary identities, then we turn to define the maps Y ,

and $\mathsf{\Lambda}$ in the quantum case.

Recall that the quantum binomial coefficient is defined by:

$$\begin{bmatrix} n \\ k \end{bmatrix} = \frac{[n]!}{[n-k]![k]!}$$

3.6.1 Lemma.

$$\begin{bmatrix} n+1 \\ k \end{bmatrix} = q^{n+1-k} \begin{bmatrix} n \\ k-1 \end{bmatrix} + q^{-k} \begin{bmatrix} n \\ k \end{bmatrix}$$

$$= q^{-n-1+k} \begin{bmatrix} n \\ k-1 \end{bmatrix} + q^k \begin{bmatrix} n \\ k \end{bmatrix}$$

Proof. We have for example,

$$q^{n+1-k} \begin{bmatrix} n \\ k-1 \end{bmatrix} + q^{-k} \begin{bmatrix} n \\ k \end{bmatrix}$$

$$= \begin{bmatrix} n+1 \\ k \end{bmatrix} \left(\frac{q^{n+1-k}[k]}{[n+1]} + \frac{q^{-k}[n-k+1]}{[n+1]} \right) = \begin{bmatrix} n+1 \\ k \end{bmatrix}.$$

$\square$

3.6.2 Lemma.

$$\sum_{\substack{R \subset \{1,\ldots,n\} \\ |R| = k}} q^{-2t_n(R)} = q^{-k(n-k)} \begin{bmatrix} n \\ k \end{bmatrix}$$

$$\sum_{\substack{R \subset \{1,\ldots,n\} \\ |R| = k}} q^{2t_n(R)} = q^{k(n-k)} \begin{bmatrix} n \\ k \end{bmatrix}$$

where $t_n(R)$ is the minimal number of transpositions that it takes to move the subset R to the subset $\{n+1-k,\ldots,n\}$.

Proof. The proof will follow by induction. We indicate the proof of the first formula; the second follows similarly.

$$\sum_{\substack{R \subset \{1,\ldots,n+1\} \\ |R| = k}} q^{-2t_{n+1}(R)}$$

$$= \sum_{\substack{R \subset \{1,\ldots,n+1\} \\ |R| = k \\ n+1 \in R}} q^{-2t_{n+1}(R)} + \sum_{\substack{R \subset \{1,\ldots,n+1\} \\ |R| = k \\ n+1 \notin R}} q^{-2t_{n+1}(R)}$$

$$= q^{-(k-1)(n-k+1)} \begin{bmatrix} n \\ k-1 \end{bmatrix} + q^{-2k} q^{-k(n-k)} \begin{bmatrix} n \\ k \end{bmatrix}.$$

Apply Lemma 3.6.1 to complete the proof. $\square$

3.6.3 Definition. Let $\overset{1}{\cup}=\overset{1}{\cup}_A= \cup_A : \mathbf{C} \to V_A^{1/2} \otimes V_A^{1/2}$. Having defined $\overset{n-1}{\cup}=\overset{n-1}{\cup}_A: \mathbf{C} \to (V_A^{1/2})^{\otimes 2(n-1)}$, define $\overset{n}{\cup}=\overset{n}{\cup}_A$ to be the composition

$$\overset{n}{\cup}: \mathbf{C} \overset{\overset{n-1}{\cup}}{\to} \underbrace{V_A^{1/2} \otimes \ldots \otimes V_A^{1/2}}_{2(n-1)} \overset{=}{\to}$$

$$\underbrace{V_A^{1/2} \otimes \ldots \otimes V_A^{1/2}}_{(n-1)} \otimes \mathbf{C} \otimes \underbrace{V_A^{1/2} \otimes \ldots \otimes V_A^{1/2}}_{(n-1)} \overset{1 \otimes \cup \otimes 1}{\to} (V_A^{1/2})^{\otimes 2n}.$$

The map $\overset{n}{\cap}=\overset{n}{\cap}_A$ is defined dually, and is also in analogy with the classical case.

3.6.4 Lemma.

$$\overset{n}{\cup}_A (1) = \sum_{S \subset \{1,\ldots,n\}} (iA)^{n-2|S|} x_1^S \otimes \cdots \otimes x_n^S \otimes \bar{x}_n^S \otimes \cdots \otimes \bar{x}_1^S$$

where

$$x_m^S = \begin{cases} y & \text{if } m \in S \\ x & \text{if } m \notin S \end{cases},$$

and

$$\bar{x}_m^S = \begin{cases} x & \text{if } m \in S \\ y & \text{if } m \notin S \end{cases}$$

Also we have,

$$\overset{n}{\cap}_A ((x_1 \otimes \cdots \otimes x_n) \otimes (\bar{x}_n \otimes \cdots \otimes \bar{x}_1))$$

$$= \begin{cases} 0 & \text{if } x_k = \bar{x}_k \text{ for some } k = 1,\ldots,n \\ (iA)^{n-2s} & \text{if } \{x_k, \bar{x}_k\} = \{x,y\} \text{ for } k = 1,\ldots,n \end{cases}$$

where $x_k, \bar{x}_k \in \{x,y\}$ for all $k = 1, 2, \ldots, n$ and $s = \#\{k : x_k = y\}$.

Proof. As in the classical case, the proof follows by induction (See Lemmas 2.7.3 and 2.5.3). In case $n = 2$ the non-zero values are as follows:

$$\overset{2}{\cap} (x \otimes x \otimes y \otimes y) = (iA)^2,$$

$$\overset{2}{\cap} (x \otimes y \otimes x \otimes y) = 1,$$

$$\overset{2}{\cap} (y \otimes x \otimes y \otimes x) = 1,$$

$$\overset{2}{\cap} (y \otimes y \otimes x \otimes x) = (iA)^{-2}.$$

$\square$

3.6.5 Definition. Let $j \in \{0, 1/2, 1, 3/2, \ldots\}$. Define a map $\mu_j : (V_A^{1/2})^{\otimes 2j} \to V_A^j$ via

$$\mu_j(x_1 \otimes x_2 \otimes \cdots \otimes x_{2j}) = x_1 \cdot x_2 \cdot \ \cdots \ \cdot x_{2j}$$

where the multiplication on the right occurs in the ring of polynomials in non-commuting variables x, y. Evidently, μ_j is a $U_q(sl(2))$ map and its kernel is L_A^{2j}.

Next we mimic the classical case to define, for (a, b, j) an admissible triple of half-integers, a $U_q(sl(2))$ invariant map

$$\overset{ab}{\underset{j}{Y}} : (V_A^{1/2})^{\otimes 2j} \to (V_A^{1/2})^{\otimes 2a} \otimes (V_A^{1/2})^{\otimes 2b}$$

via the formula

$$\overset{ab}{\underset{j}{Y}} = (\overset{A}{+_{2a}} \otimes \overset{A}{+_{2b}}) \circ (\ |_{a+j-b} \otimes \overset{a+b-j}{U_A} \otimes \ |_{b+j-a}\ ) \circ \overset{A}{+_{2j}}$$

where $|_m$ is the identity map on m tensor factors of $V_A^{1/2}$ (cf. Section 2.5.4).

3.6.6 Theorem. *(1) Let $A \in \mathbf{C}$. Let (a, b, j) denote an admissible triple such that if A is a primitive $4r$th root of unity, then*

$$\max\{2a, 2b, 2j\} < r. \text{ Then the } U_q(sl(2)) \text{ invariant map } \bigvee_{j}^{ab} :$$

$(V_A^{1/2})^{\otimes 2j} \to (V_A^{1/2})^{\otimes 2a} \otimes (V_A^{1/2})^{\otimes 2b}$ *is defined and*

$$\mu_a \otimes \mu_b \left(\bigvee_{j}^{ab} (\phi_j(e_{j,j})) \right) =$$

$$= \sum_{u+v=j} i^{(b-v)-(a-u)} A^{(b-v)(b+v+1)-(a-u)(a+u+1)} \begin{bmatrix} a+b-j \\ a-u \end{bmatrix} e_{a,u} \otimes e_{b,v}.$$

(2) Let $A \in \mathbf{C}$. Choose $a, b \in \{0, 1/2, 1, 3/2, \ldots\}$ so that if A is a primitive $4r$th root of unity, then $2a + 2b < r$. There is a direct sum decomposition

$$V_A^a \otimes V_A^b = \bigoplus_j \mu_a \otimes \mu_b \left(\bigvee_{j}^{ab} \left(\phi_j \left(V_A^j \right) \right) \right)$$

where the sum is taken over all j such that (a, b, j) is admissible. Furthermore, if (a, b, j) is admissible, then any $U_q(sl(2))$ invariant

$$\text{map } V_A^j \to V_A^a \otimes V_A^b \text{ is a scalar multiple of } \mu_a \otimes \mu_b \circ \bigvee_{j}^{ab} \circ \phi_j.$$

Proof. Recall that $e_{j,t}$ denotes the weight vector in V_A^j of weight t, and that $\omega_j(e_{j,t}) = A^{(j-m)(j+m)} x^{j+m} y^{j-m}$. The restrictions on

$a, b,$ and j insure that $\bigvee_{j}^{ab}$ is defined. The formula for $\mu_a \otimes$

$\mu_b \left(\bigvee_{j}^{ab} (\phi_j(e_{j,j})) \right)$ follows by computation using Lemma 3.6.4, Lemma 3.6.2, and Proposition 3.5.7.

For part (2), the restriction on a and b insures that $\bigvee_{j}^{ab}$ is defined and non-zero for all j such that (a, b, j) is admissible and that the representations V_A^j are irreducible since $j \leq a + b < r/2$.

The argument that $V_A^a \otimes V_A^b$ splits as a direct sum of the images of the V_A^j follows the same lines as in the proof of Theorem 2.5.5. $\square$

3.6.7 Definition. There is a $U_q(sl(2))$ invariant map $\bigwedge\limits_{ab}^{j}$: $(V_A^{1/2})^{\otimes 2a} \otimes (V_A^{1/2})^{\otimes 2b} \to (V_A^{1/2})^{\otimes 2j}$, defined for admissible triples (a, b, j) as follows:

$$\bigwedge_{ab}^{j} = {\overset{A}{+}}_{2j} \circ \left(|_{a+j-b} \otimes {\overset{a+b-j}{\cap}}_A \otimes |_{j+b-a} \right) \circ \left({\overset{A}{+}}_{2a} \otimes {\overset{A}{+}}_{2b} \right).$$

The composition $\mu_j \circ \bigwedge\limits_{ab}^{j} \circ (\phi_a \otimes \phi_b) : V_A^a \otimes V_A^b \to V_A^j$ is also $U_q(sl(2))$ invariant, and it corresponds to the projection of $V_A^a \otimes V_A^b$ onto the direct summand that is isomorphic to V_A^j.

3.6.8 Lemma. *For $u + v = j$, (a, b, j) admissible, and $\max\{2a, 2b, 2j\} < r$ if A is a primitive $4r$th root of unity, we have*

$$\left(\mu_j \circ \bigwedge_{ab}^{j} \circ (\phi_a \otimes \phi_b) \right) (e_{a,u} \otimes e_{b,v}) =$$

$$= i^{(b-v)-(a-u)} A^{(b-v)(b+v+1)-(a-u)(a+u+1)} \frac{[a+b-j]![a+u]![b+v]!}{[2a]![2b]!} e_{j,j}.$$

Proof. The computation is analogous to the proof of Lemma 2.7.4, but relies on Lemma 3.6.2 and Proposition 3.5.7 to take care of the powers of A. $\square$

3.6.9 The quantum Clebsch-Gordan coefficients.

We have maps $\mu_a \otimes \mu_b \circ \bigvee\limits_{j}^{ab} \circ \phi_j : V_A^j \to V_A^a \otimes V_A^b$ when the triple (a, b, j) is admissible. Define the *quantum Clebsch-Gordan coefficient* $C_{u,v,t}^{a,b,j}$ to be the coefficient in the sum

$$\mu_a \otimes \mu_b \left(\bigvee_{j}^{ab} (\phi_j(e_{j,t})) \right) = \sum_{u+v=t} C_{u,v,t}^{a,b,j} e_{a,u} \otimes e_{b,v}.$$

3.6.10 Lemma. *The quantum Clebsch-Gordan coefficients satisfy the following recursion relation*

$$[j+t+1]C_{u,v,t}^{a,b,j} = [a+u+1]A^{2v}C_{u+1,v,t+1}^{a,b,j}+[b+v+1]A^{-2u}C_{u,v+1,t+1}^{a,b,j}.$$

Hence,

$$C_{u,v,t}^{a,b,j}$$

$$= i^{(b-v)-(a-u)}A^{(b-v)(b+v+1)-(a-u)(a+u+1)}\frac{[j+t]![j-t]![a+b-j]!}{[2j]![a+u]![b+v]!}.$$

$$\left(\sum_{z,w:\ z+w=j-t}(-1)^z A^{(z-w)(j+t+1)}\frac{[a+u+z]![b+v+w]!}{[z]![w]![a-u-z]![b-v-w]!}\right).$$

The sum is understood to be over all integers z, w such that $z+w = j - t$ and all the factorials are of non-negative integers. Furthermore, $u, v,$ and t are weights of $V_A^a, V_A^b,$ and $V_A^j,$ respectively, and $u + v = t$.

Proof. As in the classical case, the recursion relation is found by applying F to the equation that defines the Clebsch-Gordan coefficient. The closed form is determined by solving the recursion using the value

$$C_{u,v,j}^{a,b,j} = i^{(b-v)-(a-u)}A^{(b-v)(b+v+1)-(a-u)(a+u+1)}\frac{[a+b-j]!}{[a-u]![b-v]!}.$$

□

3.7 Quantum network evaluation. Recall that in the classical case the key computations were the evaluations of the closed "theta-nets" and the closed "tetrahedral" networks. In [18], these network evaluations are given in case $A = e^{i\pi/(2r)}$ and their diagrammatic computations can be applied to the case of generic values of A as well. We use the machinery developed above to give alternative computations.

3.7.1 Theorem. *Let $A \in \mathbb{C}$. Choose $a, b, j, k \in \{0, 1/2, 1, 3/2, \ldots\}$ such that*

     1. (a, b, k) and (a, b, j) are admissible triples;

     2. If A is a primitive $4r$th root of unity, then assume that $a, b, j, k < r/2$.

Then

$$\mu_k \left(\bigwedge_{ab}^{k} \left(\bigvee_{j}^{ab} (\phi_j (e_{j,j})) \right) \right) = \left[\Theta(a, b, j) \delta_j^k / \Delta_j \right] e_{j,j}$$

where δ_j^k is a Kronecker δ function, $\Delta_j = (-1)^{2j}[2j + 1]$, and

$$\Theta(a, b, k) =$$

$$(-1)^{a+b+k} \frac{[a + b - k]![a - b + k]![-a + b + k]![a + b + k + 1]!}{[2a]![2b]![2k]!}.$$

3.7.2 Remark. In case $k = j = 0$ and $a = b$, the formula reduces to $\Delta_a = (-1)^{2a}[2a + 1]$ and this is the value of a closed loop with a single projector attached. In a similar fashion, the value $\Theta(a, b, k)$ is the value of a closed network as in the paragraph preceding Corollary 2.7.7.

Proof. By the assumption on A, the representations V_A^j and V_A^k are irreducible. Therefore, the given composition is 0 when $k \neq j$, and when $k = j$ it is a multiple of the identity.

In Theorem 3.6.6 we computed the value of $\bigvee$ on a highest weight vector, and in Lemma 3.6.8 we computed $\bigwedge$. We combine these results and manipulate the expression for the exponent of A while recalling that $A^2 = q$ and obtain the following:

$$\mu_j\left(\bigwedge_{ab}^{j}\left(\bigvee_{j}^{ab}(\phi_j(e_{j,j}))\right)\right)$$

$$= \mu_j\left(\bigwedge_{ab}^{j}\left(\sum_{u+v=j} i^{(b-v)-(a-u)} A^{(b-v)(b+v+1)-(a-u)(a+u+1)}\right.\right.$$

$$\left.\left.\begin{bmatrix} a+b-j \\ a-u \end{bmatrix}\phi_a(e_{a,u})\otimes\phi_b(e_{b,v})\right)\right)$$

$$= e_{j,j}\sum_{u+v=j}(-1)^{(b-v)-(a-u)}q^{(b-v)(b+v+1)-(a-u)(a+u+1)}.$$

$$\begin{bmatrix} a+b-j \\ a-u \end{bmatrix}\frac{[a+b-j]![a+u]![b+v]!}{[2a]![2b]!}$$

$$= e_{j,j}(-1)^{a+b-j}\frac{([a+b-j]!)^2}{[2a]![2b]!}\sum_{u+v=j}\frac{q^{(b-v)(b+v+1)}}{q^{(a-u)(a+u+1)}}\frac{[a+u]![b+v]!}{[a-u]![b-v]!}$$

$$= e_{j,j}(-1)^{a+b-j}\frac{[a+b-j]![a+j-b]![b+j-a]![b+j+a+1]!}{[2a]![2b]![2j+1]!}$$

The last equality is a quantum combinatoric identity that is analogous to the identity used in the proof of Theorem 2.7.6. Thus to complete the proof we need use the following:

3.7.3 Lemma.

$$\sum_{u+v=j}\frac{q^{(b-v)(b+v+1)}}{q^{(a-u)(a+u+1)}}\frac{[a+u]![b+v]!}{[a-u]![b-v]!}$$

$$= \begin{bmatrix} a+b+j+1 \\ 2j+1 \end{bmatrix}[a+j-b]![b+j-a]!$$

Proof. The proof is a quantization of the argument that we gave during the proof of Theorem 2.7.6. We rely on the quantum combinatorial identities stated in Lemma 3.6.2. Furthermore, we recall from the proof of Proposition 3.5.4 that

$$\sum_{\sigma\in\Sigma_n}q^{\pm2T(\sigma)} = A^{\pm n(n-1)}[n]!$$

where $T(\sigma)$ is the minimal number of transpositions of the form $(k, k+1)$ that it takes to write σ.

Consider a sequence $0 = a_0 < a_1 < \ldots < a_r = n$, and let $f : \{1, \ldots, n\} \to \{1, \ldots, n\}$ denote a bijection for which the restrictions $f|_{\{a_p + 1, \ldots, a_{p+1}\}}$ are increasing for $0 \le p \le r - 1$. Thus when $r = 2$, the permutation f acts like a shuffle to a deck of cards. An arbitrary permutation σ_p of the set $\{a_p + 1, \ldots, a_{p+1}\}$, is extended to a permutation of $\{1, \ldots, n\}$ by the identity. Then compute the number of adjacent transpositions that it takes to write $f \circ (\sigma_0 \sigma_1 \cdots \sigma_{r-1})$ to obtain

$$T(f \circ (\sigma_0 \sigma_1 \cdots \sigma_{r-1})) = T(f) + \sum_{p=0}^{r-1} T(\sigma_p).$$

(The preceding formula follows rather easily from the depiction of permutations as strings crossing in the plane.)

In a similar fashion if $f^{-1}|_{\{a_p + 1, \ldots, a_{p+1}\}}$ is increasing for $0 \le p \le r - 1$, then

$$T((\sigma_0 \sigma_1 \cdots \sigma_{r-1}) \circ f) = T(f) + \sum_{p=0}^{r-1} T(\sigma_p).$$

If $f : \{1, \ldots, n\} \to \{1, \ldots, n\}$ is a bijection such that the restrictions of f to $\{1, \ldots, m\}$ and to $\{m + 1 \ldots, n\}$ are both increasing, then

$$T(f) = t_n(f(\{m + 1, \ldots, n\}))$$

because $t_n(f(\{m + 1, \ldots, n\}))$ is the minimal number of transpositions that is needed to put $f(\{m + 1, \ldots, n\})$ back into its standard position.

Consider the set, B, of bijections $f : \{1, 2, \ldots, a+b+j+1\} \to \{1, 2, \ldots, a+b+j+1\}$ such that the value $f(a+j-b+1)$ is greater than every element of $f(\{1, \ldots, a + j - b\})$ and less than every element of $f(\{a + j - b + 2, \ldots, 2j + 1\})$. We evaluate $\sum_{\sigma \in B} q^{2T(\sigma)}$ in two ways to obtain the desired identity.

On the one hand, when $S \subset \{1, \ldots, a + b + j + 1\}$ is a subset of size $|S| = a + b - j$, let

$$B_S = \{\sigma \in B : \sigma(\{2j + 2, \ldots, a + b + j + 1\}) = S\}.$$

For such an S, let f_S be the unique permutation of $\{1, \ldots, a + b + j + 1\}$ such that $f_S \in B_S$ and the restrictions $f|_{\{1, \ldots, 2j + 1\}}$ and $f|_{\{2j + 2, \ldots, a + b + j + 1\}}$ are both increasing.

The set B can be written as the disjoint union of the sets B_S as S ranges over all subsets of size $a + b - j$. Furthermore,

$$B_S = \{f_S \circ (\sigma_1 \sigma_2 \sigma_3) : \sigma_1 \in \Sigma_{\{1, \ldots, a + j - b\}},$$

$$\sigma_2 \in \Sigma_{\{a + j - b + 2, \ldots, 2j + 1\}}, \ \& \ \sigma_3 \in \Sigma_{\{2j + 2, \ldots, a + b + j + 1\}}\}.$$

From this formulation we have,

$$\sum_{\sigma \in B_S} q^{2T(\sigma)} = q^{2t(S)} \left(\sum_{\sigma_1} q^{2T(\sigma_1)} \right) \left(\sum_{\sigma_2} q^{2T(\sigma_2)} \right) \left(\sum_{\sigma_3} q^{2T(\sigma_3)} \right)$$

$$= A^N q^{2t(S)} [a + j - b]! [b + j - a]! [a + b - j]!$$

where the exponent is given by

$$N = (a+j-b)(a+j-b-1) + (b+j-a)(b+j-a-1) + (a+b-j)(a+b-j-1).$$

Next we sum over all the subsets S to obtain

$$\sum_{\sigma \in B} q^{2T(\sigma)} = \sum_{\substack{S \subset \{1, \ldots, a + b + j + 1\} \\ |S| = a + b - j}} \ \sum_{\sigma \in B_S} q^{2T(\sigma)}$$

$$= A^N q^{(a+b-j)(2j+1)} \begin{bmatrix} a+b+j+1 \\ 2j+1 \end{bmatrix} [a+j-b]![b+j-a]![a+b-j]!.$$

On the other hand, we compute $\sum_{\sigma \in B} q^{2T(\sigma)}$ by examining the images of the various σ. Specifically, for $a + j - b \leq \ell \leq 2a$, let

$$B_\ell = \{\sigma \in B : \sigma(a+j-b+1) = \ell+1\}.$$

Then B is the disjoint union of the B_ℓs.

For $a + j - b \leq \ell \leq 2a$, and for $S \subset \{2j+2, \ldots, a+b+j+1\}$ of size $|S| = \ell - (a+j-b)$, let

$$B_{\ell,S} = \{\sigma \in B_\ell : \sigma^{-1}(\{1, \ldots, \ell\}) = \{1, \ldots, a+j-b\} \cup S\}.$$

Let $g_{\ell,S}$ be the unique bijection on $\{1, 2, \ldots, a+b+j+1\}$ such that $g_{\ell,S} \in B_{\ell,S}$ and the restrictions of $g_{\ell,S}^{-1}$ to each of $\{1, \ldots, \ell\}$ and $\{\ell+2, \ldots, a+b+j+1\}$ are both increasing. Then B_ℓ is the disjoint union of the $B_{\ell,S}$, and

$$B_{\ell,S} = \{(\sigma_1 \sigma_2) \circ g_{\ell,S} : \sigma_1 \in \Sigma_{\{1,\ldots,\ell\}} \ \& \ \sigma_2 \in \Sigma_{\{\ell+2\ldots,a+b+j+1\}}\}.$$

As before, we examine the number of adjacent transpositions required to write various permutations, and obtain

$$\sum_{\sigma \in B_{\ell,S}} q^{2T(\sigma)} = \left(q^{2T(g_{\ell,S})}\right) \left(\sum_{\sigma_1} q^{2T(\sigma_1)}\right) \left(\sum_{\sigma_2} q^{2T(\sigma_2)}\right)$$

$$= A^M q^{2T(g_{\ell,S})} [\ell]![a+b+j-\ell]!$$

where the exponent $M = \ell(\ell-1) + (a+b+j-\ell)(a+b+j-\ell-1)$.

Observe that $T(g_{\ell,S})$ is the number of transpositions of the form $(k, k+1)$ that are needed to move $\{a+j-b+1, \ldots, \ell\}$ into the set S. This is the number of transpositions required to move $\{a+j-b+1, \ldots, \ell\}$ into the final position in the interval $\{1, \ldots, a+b+j+1\}$, which is $|S|(a+b+j+1-\ell)$, minus the number required to move S into the final position, *i.e.* $t(S)$. Recapitulating,

$$T(g_{\ell,S}) = (\ell - a - j + b)(a + b + j + 1 - \ell) - t(S).$$

We recall from Lemma 3.6.2 that

$$\sum_{\substack{S \subset \{2j+2,\ldots,a+b+j+1\} \\ |S| = \ell - a - j + b}} q^{-2t(S)}$$

$$= q^{-(\ell-a-j+b)(2a-\ell)} \begin{bmatrix} a+b-j \\ \ell-a-j+b \end{bmatrix}.$$

Putting everything together, we get

$$\sum_{\sigma \in B} q^{2T(\sigma)} = \sum_{\ell} \left(\sum_S \sum_{\sigma \in B_{\ell,S}} q^{2T(\sigma)} \right)$$

$$= \sum_{\ell} A^E \begin{bmatrix} a+b-j \\ \ell-a-j+b \end{bmatrix} [\ell]![a+b+j-\ell]!$$

$$= A^P \sum_{\ell} q^{2\ell(j+1)} \begin{bmatrix} a+b-j \\ \ell-a-j+b \end{bmatrix} [\ell]![a+b+j-\ell]!$$

where the exponent E is given as

$$E = M + 4(\ell - a - j + b)(a + b + j + 1 - \ell) - 2(\ell - a - j + b)(2a - \ell)$$

and the related exponent P is the following

$$P = (a + b + j)(a + b + j - 1) - 4(a + j - b)(b + j + 1).$$

Set $\ell = a+u$ and set $j = u+v$, and compare the two expressions for $\sum_{\sigma \in B} q^{2T(\sigma)}$. We get

$$\sum_{u+v=j} q^{2(a+u)(j+1)} \frac{[a+u]![b+v]!}{[a-u]![b-v]!} =$$

$$= A^{N-P} q^{(a+b-j)(2j+1)} \begin{bmatrix} a+b+j+1 \\ 2j+1 \end{bmatrix} [a+j-b]! [b+j-a]!$$

Lemma 3.7.3 follows by manipulating the exponents. This completes the proof of Theorem 3.7.1. □

3.7.4 Remark. In case A is a primitive 4rth root of unity, $\Theta(a,b,j) \neq 0$ precisely when $a+b+j \leq (r-2)$. We will give the representation theoretic meaning in Section 4.

3.8 The quantum $6j$-symbols — generic case. In this section we define the quantum $6j$-symbols in case A is not a root of unity, and verify that the Elliott-Biedenharn identity and the orthogonality identity hold here. In Lemma 3.10.9 we establish the diagrammatic rules that are necessary to prove the analogues of the identities given in Theorem 2.7.14. We examine the symmetries of these quantum $6j$-symbols as well. In this way, we will have established that the Kauffman-Lins [18] diagrammatics apply directly to the theory of $U_q(sl(2))$ as presented in [19]. This discussion will complete the current section. In the section that follows, the quantum trace and the root of unity case are taken up in detail.

3.8.1 Definitions. Suppose that $q^r \neq 1$ for $r = 1, 2, \ldots$. We construct the space of $U_q(sl(2))$ invariant maps $V_A^k \to V_A^a \otimes V_A^b \otimes V_A^c$ in two different ways, as we did in the classical case. First, consider the composition

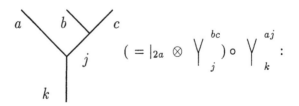

$$(V_A^{1/2})^{\otimes 2k} \rightarrow (V_A^{1/2})^{\otimes 2a} \otimes (V_A^{1/2})^{\otimes 2b} \otimes (V_A^{1/2})^{\otimes 2c}$$

for various values of j. Second, consider the composition

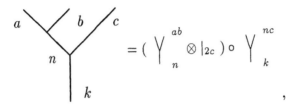

for various values of n.

The values of j and n are restricted so that (b, c, j), (a, j, k), (a, b, n), and (n, c, k) all form admissible triples. Alternatively, if one of these triples is not admissible, then we declare the corresponding map $\curlyvee$ to be the zero map.

3.8.2 Lemma. *The sets*

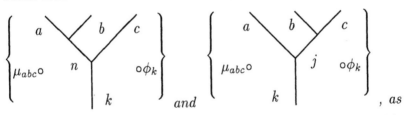

the indices n and j range in such a way that (b, c, j), (a, j, k), (a, b, n), and (n, c, k) form admissible triples, form bases for the vector space of $U_q(sl(2))$ invariant linear maps $V_A^k \rightarrow V_A^a \otimes V_A^b \otimes V_A^c$ provided that A is not a root of unity. Here $\mu_{abc} = \mu_a \otimes \mu_b \otimes \mu_c$ is the tensor product of the multiplication maps.

Proof. The proof follows as in the classical case, using Theorem 3.6.6. □

3.8.3 Definition.   Define the *quantum 6j-symbol* to be the

coefficient $\left\{ \begin{array}{ccc} a & b & n \\ c & k & j \end{array} \right\}_q$ in the following equation.

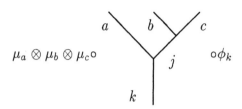

$= \sum_n \left\{ \begin{array}{ccc} a & b & n \\ c & k & j \end{array} \right\}_q$ $\mu_a \otimes \mu_b \otimes \mu_c \circ$

Thus, the trees are used to define bases for $\text{hom}_{U_q(sl(2))}(V^k, V^a \otimes V^b \otimes V^c)$ and the quantum $6j$-symbol is the change of basis matrix. By convention, $\left\{ \begin{array}{ccc} a & b & n \\ c & k & j \end{array} \right\}_q = 0$ if any of the triples (b, c, j), (a, j, k), (a, b, n), (n, c, k) is not admissible.

For example, consider the case when $a = b = c = k = 1/2$. One can compute directly from the definitions that

$$\left\{ \begin{array}{ccc} 1/2 & 1/2 & 0 \\ 1/2 & 1/2 & 0 \end{array} \right\}_q = -1/[2],$$

$$\left\{ \begin{array}{ccc} 1/2 & 1/2 & 1 \\ 1/2 & 1/2 & 0 \end{array} \right\}_q = 1,$$

$$\left\{ \begin{array}{ccc} 1/2 & 1/2 & 0 \\ 1/2 & 1/2 & 1 \end{array} \right\}_q = [3]/[2]^2,$$

and

$$\left\{ \begin{array}{ccc} 1/2 & 1/2 & 1 \\ 1/2 & 1/2 & 1 \end{array} \right\}_q = 1/[2].$$

Observe that these calculations are analogous to the classical case. Moreover, the recursive method given in Section 2.8 that is used to compute the values of the $6j$-symbols in general also applies to the quantum case.

3.8.4 Identities among diagrams. The key to simplifying the computations of the classical case was to derive identities among the diagrams that represented various maps. To this end we observe that Theorem 3.7.1 gives the evaluation of the closed theta network for certain values of (a, b, j).

The identities that are expressed in Lemma 2.6.4 and their proofs go over to the quantum case as they are stated. The essence of the proof is contained in the fact that

$$\left(\begin{array}{cc} 0 & iA \\ (iA)^{-1} & 0 \end{array} \right)^2 = \left(\begin{array}{cc} 1 & 0 \\ 0 & 1 \end{array} \right).$$

The matrix on the left represents the maps $\cup$ and $\cap$, and this element squared represents the composition of the operators $\cap \otimes | \circ | \otimes \cup$.

3.8.5 Theorem (Orthogonality). *Suppose that $q^r \neq 1$. The quantum $6j$-symbols satisfy the following relation:*

$$\sum_j \left\{ \begin{array}{ccc} b & c & j \\ k & a & n \end{array} \right\}_q \left\{ \begin{array}{ccc} a & b & m \\ c & k & j \end{array} \right\}_q = \delta_{m,n}.$$

Proof. Identical to the proof of Theorem 2.6.6. $\square$

3.8.6 Theorem (Elliott-Biedenharn Identity). *The following relation holds among the $6j$-symbols when $q^r \neq 1$.*

$$\left\{ \begin{array}{ccc} c & d & h \\ g & e & f \end{array} \right\}_q \cdot \left\{ \begin{array}{ccc} b & h & k \\ g & a & e \end{array} \right\}_q$$

$$= \sum_j \left\{ \begin{array}{ccc} b & c & j \\ f & a & e \end{array} \right\}_q \cdot \left\{ \begin{array}{ccc} j & d & k \\ g & a & f \end{array} \right\}_q \cdot \left\{ \begin{array}{ccc} c & d & h \\ k & b & j \end{array} \right\}_q.$$

Proof. Identical to the proof of 2.6.7. □.

3.9 Diagrammatics of weight vectors (quantum case).

In the classical (unquantized) case we represented the weight vectors x and y by white and black vertices, respectively. Similar diagrams apply in the quantum case as well, and manipulation of these diagrams can be used to provide proofs. The next two lemmas are useful for such manipulations even if we do not use them in full force. In particular, you may find some computations run more smoothly for you by using a variety of these facts.

3.9.1 Lemma.   *The following identities hold.*

Proof. In this notation,

$$\cup = iA \, \mathbf{\delta} \, \mathbf{\delta} + (iA)^{-1} \, \mathbf{\delta} \, \mathbf{\delta},$$

while

$$\cap = iA \, \Upsilon \, \Upsilon + (iA)^{-1} \, \Upsilon \, \Upsilon.$$

The proof follows easily. □

3.9.2 Lemma. *The following identities hold.*

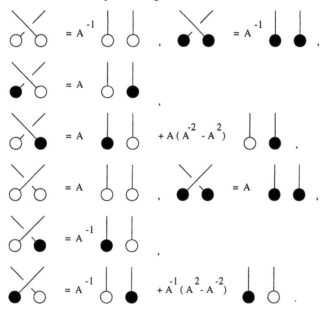

Proof. Use the bracket relation. □

3.10 Twisting rules. The goal of the current section is to prove the twisting rules that are established in Lemma 3.10.9. Our proof closely follows the one presented in [18]. The Lemma will be used in turn to establish Theorem 3.12 which contains the quantum analogues of the results in Theorem 2.7.14. The relationships established there represent many of the important identities that hold for the quantum $6j$-symbols in the generic case (cf. [19]). First we establish some notation and basic results.

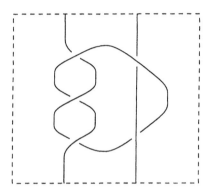

Figure 6: A tangle diagram

3.10.1 Definition. A *tangle diagram* is a diagram of a collection of knotted and linked proper arcs and circles in a rectangular box in which the arcs in the diagram terminate on the top and bottom rectangular faces of the box. The box is projected to the plane of the paper, and the diagram is assumed to be in general position meaning that at most two arcs cross at a given point, and such a crossing is transverse. The diagram depicts over and under crossing information in the standard way: at a crossing point, the arc that is farthest from the plane of projection is broken. An example of a tangle diagram is given in Figure 6.

A tangle diagram for which n arcs come out of the bottom of the diagram and m arcs come out of the top of the diagram represents a map

$$T : (V_A^{1/2})^{\otimes n} \to (V_A^{1/2})^{\otimes m}$$

via the association of maps $\cap$ and $\cup$ to generic maximal and minimal points and via the association of the bracket identity to each crossing. In the sequel, we will identify the diagram and the represented map.

Two tangle diagrams are *regularly isotopic* if one can be obtained from the other by a sequence of the type II and type III Reidemeister moves. More precisely, when a height function is chosen on the plane of projection, then the 3-dimensional analogues of the identities expressed in Lemma 2.3.2 together with the diagrammatic representations of the identities $s_j s_j^{-1} = 1$, $s_j s_{j+1} s_j = s_{j+1} s_j s_{j+1}$, and their obvious variants obtained by changing appropriate crossings are the moves that generate regular isotopy of tangle diagrams. These last two moves are depicted below. Thus *a regular isotopy* is a sequence of these diagrammatic moves, and each such move corresponds to an identity between maps represented by the respective diagrams.

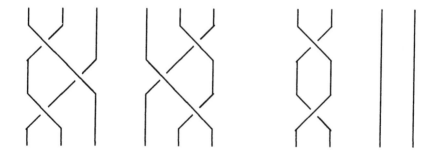

We have established the following:

3.10.2 Lemma. *(cf. Lemma 2.7.2). Regularly isotopic tangle diagrams with n strings at the bottom and m strings at the top represent the same map*

$$T : (V_A^{1/2})^{\otimes n} \to (V_A^{1/2})^{\otimes m}.$$

3.10.3 Definition.   Let $a, b \in \{0, 1/2, 1, \ldots\}$. Define maps $\chi_{ba}^{ab}(+)$ and $\chi_{ba}^{ab}(-)$ that are quantizations of the map that switches factors defined in Lemma 2.7.5. Thus,

$$\chi_{ba}^{ab}(+), \chi_{ba}^{ab}(-) : (V_A^{1/2})^{2a} \otimes (V_A^{1/2})^{2b} \to (V_A^{1/2})^{2b} \otimes (V_A^{1/2})^{2a}.$$

The (unquantized) switching map χ_{ba}^{ab} can be written as a product of adjacent transpositions since it represents a permutation; choose a minimal such product. Each transposition $\sigma_k = (k, k+1)$ in the product is lifted to a braid generator $\hat{\sigma}_k = s_k$ to define $\chi_{ba}^{ab}(+)$, and it is lifted to $\hat{\sigma}_k^{-1} = s_k^{-1}$ to define $\chi_{ba}^{ab}(-)$. Hence, in the diagrammatic representation of $\chi_{ba}^{ab}(+)$ the $2a$ strings on the left cross over the $2b$ strings on the right as one reads from top to bottom; in $\chi_{ba}^{ab}(-)$ the strings on the top right cross over those on the top left.

Define maps $T_n(\pm) : (V_A^{1/2})^{\otimes 2n} \to (V_A^{1/2})^{\otimes 2n}$ that are represented by half-twists in the cable of n-strings. Specifically, the half-twists can be spelled out in terms of braid generators as the products:

$$T_n(-) = (s_1^{-1})(s_2^{-1} s_1^{-1}) \cdots (s_{n-1}^{-1} s_{n-2}^{-1} \cdots s_1^{-1})$$

while

$$T_n(+) = (s_1 \cdots s_{n-2} s_{n-1})(s_1 s_2 \cdots s_{n-3} s_{n-2}) \cdots (s_1 s_2) s_1.$$

To obtain the desired maps, $\chi_{ba}^{ab}(\pm)$ and $T_n(\pm)$ apply the bracket identity to each braid generator in the product.

3.10.4 Notation.   Suppose that S is a planar tangle with no closed loops. Thus S consists of properly embedded arcs in the rectangular region containing S. We depict S as a rectangular coupon as in the illustration on the left below. The tangle Z is a $180°$ rotation of S through a vertical axis.

Suppose that $2b$ strings come out of the top left edge of the tangle S, suppose that $2a$ strings come out of the top right of S, and $2j$ strings come out of the bottom of S. The top-right, top-left, and bottom of the tangle are called the α, β and γ regions of the boundary, respectively. Thus S represents a map $(V_A^{1/2})^{\otimes 2j} \rightarrow (V_A^{1/2})^{\otimes 2b} \otimes (V_A^{1/2})^{\otimes 2a}$. The strings that come out of the coupon are represented by the three ribbons in the picture below.

We will compare the composition $\chi_{ba}^{ab}(+) \circ S$ to the composition $(T_{2a}(-) \otimes T_{2b}(-)) \circ Z \circ T_{2j}(+)$ where each of these tangles is associated canonically with a map among tensor powers of the fundamental representation.

3.10.5 Observation. *The diagrams depicted below are isotopic but not necessarily regularly isotopic diagrams. The isotopies fix the edges of the outgoing regions.*

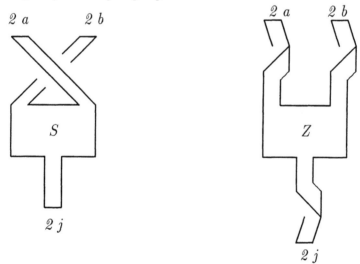

Demonstration. Cut out a piece of paper in the shape of a thick Y with fairly long edges. Arrange the edges with a crossing as in the left hand diagram and tape the ends to a table. Write

the letter S on the neighborhood of the vertex of the Y. Then rotate the vertex about the vertical axis. This gives the isotopy.
□

The maps that are represented by the diagram on the left and the diagram on the right differ by a factor of $(-A^3)^{\pm k}$, for such a factor measures the difference between isotopy and regular isotopy. We next quantify that difference.

In the planar tangle S (with β region on the upper left, α region on the upper right, and γ region on the bottom edge) suppose that x strings start and end in the β region, r strings start and end in the α region, p strings start and end in the γ region, w strings run between the β and the α region, u strings run between the β and γ regions, and t strings run between the α and the γ region. Otherwise the diagram for S can be quite arbitrary, and there is no assumption about the nesting or interlacing of these strings beyond the condition that no strings cross. The tangle Z denotes the map represented by a rotation of S by $180°$ about the vertical axis.

3.10.6 Lemma. *In the notation above*

$$\chi_{ba}^{ab}(+) \circ S = (-A^3)^{x+r+w-p}(T_{2a}(-) \otimes T_{2b}(-)) \circ Z \circ T_{2j}(+).$$

Proof. We establish some special cases then we will proceed to the general result.

Corresponding to the case $x = r = p = u = t = 0$ and $w = 2a = 2b = n$ in the notation above, we have the first equation in the following:

3.10.7 Sublemma (Case 1).

$$\chi_n(\pm) \circ \overset{n}{U} = (-A^{\pm 3})^n(T_n(\mp) \otimes T_n(\mp)) \circ \overset{n}{U}$$

and

$$\overset{n}{\cap} \circ \chi_n(\pm) = (-A^{\pm 3})^n \overset{n}{\cap} \circ(T_n(\mp) \otimes T_n(\mp))$$

where $\chi_n(\pm)$ denotes n strings crossing from top left to lower right over/under n strings running from lower left to upper right, respectively.

Proof. Bracket aficionados will recognize that

$$\chi(\pm)\cup = (-A)^{\pm 3}\cup$$

and

$$\cap\chi(\pm) = (-A)^{\pm 3}\cap,$$

with signs read respectively. (Non-aficionados are encouraged to perform this calculation.) The general result follows by induction. The trick is to regularly isotope the inner loop to an isolated region of the diagram. □

The diagrams representing $\chi_n(\pm) \overset{n}{\cup}$ and $\overset{n}{\cap} \chi_n(\pm)$ will be called *curlicues.*

For the case in which $x = r = w = a = b = 0$ and $p = n$ we have the first equation below. The second equation below corresponds to the case when all strings start and end in either the α or the β region; *i.e.* $j = 0$ and either $a = 0$ or $b = 0$.

3.10.8 Sublemma (Case 2).

$$\overset{n}{\cap} = (-A^3)^{\mp n} \overset{n}{\cap} \circ T_{2n}(\pm)$$

and

$$\overset{n}{\cup} = (-A^3)^{\pm n} T_{2n}(\mp) \circ \overset{n}{\cup}$$

where the signs are read respectively.

Proof. We induct on n. The case $n = 1$ is the same as Case 1 above when $n = 1$. For larger values of n, one small kink can be canceled from the diagram representing the map on the right (of either equation) at the expense of multiplying that factor by $(-A^3)^{\pm 1}$. Because T represents a complete half-twist, the arc from which the kink has been canceled can be isotoped away from the rest of the diagram. $\square$

We turn now to the proof of the general result. Let S' denote the tangle obtained by twisting the w curlicues out of $\chi_{ba}^{ab}(+) \circ S$. Specifically, in each string that runs between the α and β regions, the small loop is removed. These loops are removed successively. Thus we replace $\chi_w(+) \circ \cup_w$ by $(T_w(-) \otimes T_w(-)) \circ \cup_w$. Then

$$\chi_{ba}^{ab}(+) \circ S = (-A^3)^w S'$$

as in Case 1.

Similarly, let Z' denote the result of removing the curlicues from the $r, x,$ and p strings that run back and forth to the α, β, and γ regions, respectively. The nesting among arcs that run back and forth to a given region is not relevant, as can be seen by examining the figure below. In this way we obtain,

$$(T_{2a}(-) \otimes T_{2b}(-)) \circ Z \circ T_{2j}(+) = (-A^3)^{p-x-r} Z'$$

by Case 2.

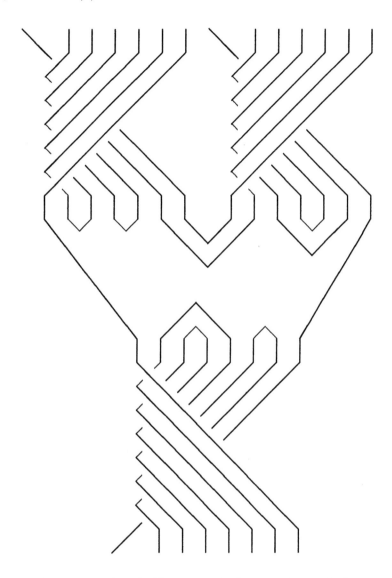

Furthermore, Z' and S' are regularly isotopic, and thus represent the same map from $(V_A^{1/2})^{\otimes 2j}$ to $(V_A^{1/2})^{\otimes 2a} \otimes (V_A^{1/2})^{\otimes 2b}$. Thus

$$\chi_{ba}^{ab}(+) \circ S = (-A^3)^w S'$$

$$= (-A^3)^w Z' = (-A^3)^{w+x+r-p}(T_{2a}(-) \otimes T_{2b}(-)) \circ Z \circ T_{2j}(+).$$

This completes the proof. $\square$

3.10.9 Lemma. *The following identities hold among the $U_q(sl(2))$ invariant maps.*

1. $\chi^{ab}_{ba}(\pm) \circ \displaystyle\bigvee_{\substack{j}}^{ba} = (-1)^{a+b-j} A^{\pm 2(a(a+1)+b(b+1)-j(j+1))} \displaystyle\bigvee_{\substack{j}}^{ab}.$

$$\underset{2j}{\overset{2a \diagdown \diagup 2b}{\diamondsuit}} = (-1)^{a+b-j} A^{2(a(a+1)+b(b+1)-j(j+1))} \underset{2j}{\overset{2a \qquad 2b}{\bigvee}}$$

2. $\displaystyle\bigwedge_{\substack{ba}}^{\;j} \circ \chi^{ba}_{ab}(\pm) = (-1)^{j-a-b} A^{\pm 2(a(a+1)+b(b+1)-j(j+1))} \displaystyle\bigwedge_{\substack{ab}}^{\;j}.$

$$\underset{2a \diagup \diagdown 2b}{\overset{\mid\, 2j}{\diamondsuit}} = (-1)^{j-a-b} A^{2(a(a+1)+b(b+1)-j(j+1))} \underset{2a \qquad 2b}{\overset{2j \mid}{\bigwedge}}$$

3.

$$\left(|_{2m} \otimes \bigvee_{\substack{j}}^{ab} \right) \circ \chi^{mj}_{jm}(\pm)$$

$$= (\chi^{ma}_{am}(\pm) \otimes |_{2b}) \circ \left(|_{2a} \otimes \chi^{mb}_{bm}(\pm) \right) \circ \left(\bigvee_{\substack{j}}^{ab} \otimes |_{2m} \right).$$

4.

$$\chi_{jm}^{mj}(\pm) \circ \left(\overset{j}{\underset{ab}{\wedge}} \otimes \mid_{2m} \right)$$

$$= \left(\mid_{2m} \otimes \overset{j}{\underset{ab}{\wedge}} \right) \circ (\chi_{am}^{ma}(\pm) \otimes \mid_{2b}) \circ \left(\mid_{2a} \otimes \chi_{bm}^{mb}(\pm) \right).$$

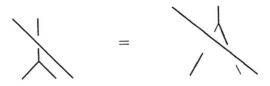

Proof. We use Lemma 3.10.6 to prove item 1 as follows. We restrict to the case of $\chi_{ba}^{ab}(+)$, the negative crossing case is similar.

The map $\overset{ba}{\underset{j}{Y}}$ can be written as a linear combination of maps S that are represented by planar tangles with no closed loops as follows. Expand each of the projectors $+_{2a}^{A}$, $+_{2b}^{A}$, and $+_{2j}^{A}$ in terms of the bracket identity. Replace any closed loop that might appear in the expansion by the factor $-A^2 - A^{-2}$. Thus we have

$$\chi_{ba}^{ab}(+) \circ \overset{ab}{\underset{j}{Y}} = \sum_S f_S(A) \chi_{ba}^{ab} S$$

where $f_S(A)$ is a polynomial in $A^{\pm 1}$ and the sum is over all planar tangles S with $2b + 2a$ outputs on the top and $2j$ outputs on the bottom. Next we consider the exponent on the factor $-A^3$ on the right hand side of 3.10.6. We have

$$x + r + w - p = a + b - j$$

regardless of the planar tangle S. This gives

$$\chi_{ba}^{ab}(+) \circ \bigvee_{j}^{ab} = \sum_{S} f_S(A)(-A^3)^{a+b-j}(T_{2a}(-) \otimes T_{2b}(-)) \circ Z \circ T_{2j}(+)$$

$$= (-A^3)^{a+b-j}(T_{2a}(-) \otimes T_{2b}(-)) \circ \bigvee_{j}^{ab} \circ T_{2j}(+)$$

because the projector $\dashuline{}_n^A$ is invariant under $180°$ rotation for $n = 2b, 2b, 2j$.

     The second step of the proof is now relatively easy. Observe that

$$T_n(-) \circ \dashuline{}_n^A = A^{n(n-1)/2} \dashuline{}_n^A.$$

This follows by applying the bracket relation to each of the negative crossings in the half-twist. In the bracket relation either straight strings or maximum and minimum points result. The optimal points occur with a coefficient of A^{-1}; and these optima are annihilated by the Jones-Wentzl projectors. Consequently, all of these terms with a coefficient of A^{-1} vanish in the bracket expansion. Similarly we compute

$$\dashuline{}_{2j}^{A} T_{2j}(+) = A^{-j(2j-1)} \dashuline{}_{2j}^{A}.$$

     Combining results we have

$$\chi_{ab}^{ab}(+) \bigvee_{j}^{ba} = (-A^3)^{a+b-j} T_{2a}(-) \otimes T_{2b}(-) \circ \bigvee_{j}^{ab} \circ T_{2j}(+)$$

$$= (-1)^{a+b-j} A^{3(a+b-j)} A^{a(2a-1)+b(2b-1)-j(2j-1)} \bigvee_{j}^{ab}$$

$$= (-1)^{a+b-j} A^{2(a(a+1)+b(b+1)-j(j+1))} \bigvee_{j}^{ab}.$$

This completes the proof of item 1.

Item 2 follows from item 1 by techniques of regular isotopy.

Items 3 and 4 follow trivially from the invariance of the map represented by regularly isotopic diagrams. □

3.11 Symmetries. Here the twisting rules will be used to adjust the $6j$-symbol to one that has full tetrahedral symmetry.

3.11.1 Lemma. *The quantum $6j$-symbols possess the following symmetry*

$$\left\{ \begin{array}{ccc} m & p & u \\ t & s & r \end{array} \right\}_q = \left\{ \begin{array}{ccc} t & s & u \\ m & p & r \end{array} \right\}_q$$

Proof. Identical to Lemma 2.7.8. □

3.11.2 Lemma.

$$\frac{\Theta(s,t,k)}{\Delta_k} \left\{ \begin{array}{ccc} m & p & k \\ t & s & r \end{array} \right\}_q = \frac{\Theta(m,r,s)}{\Delta_m} \left\{ \begin{array}{ccc} p & k & m \\ s & r & t \end{array} \right\}_q \quad (3)$$

$$= \frac{\Theta(r,t,p)}{\Delta_p} \left\{ \begin{array}{ccc} k & m & p \\ r & t & s \end{array} \right\}_q \quad (4)$$

where the functions $\Theta(-,-,-)$ and Δ are defined as in Theorem 3.7.1.

Proof. Same as the proof of 2.7.9. □

3.11.3 Lemma. *The symbol,*

$$\left[\begin{array}{ccc} a & b & f \\ e & d & c \end{array} \right]_q =$$

$$\mathrm{TET}(a,b,c,d,e,f)_q / (\sqrt{\Theta(a,b,f)}\sqrt{\Theta(d,e,f)}\sqrt{\Theta(a,c,d)}\sqrt{\Theta(b,c,e)})$$

where a choice of each square root is made once and for all, is invariant under all permutations of its columns and under the exchange of any pair of elements in the top row with the corresponding pair in the bottom row. Equivalently, the symbol $\begin{bmatrix} a & b & f \\ e & d & c \end{bmatrix}_q$ *is invariant under the permutations of the set*

$$\{\{a, b, f\}, \{a, c, d\}, \{b, c, e\}, \{d, e, f\}\}.$$

Proof. The proof of Lemma 2.7.12 needs to be modified as follows. When the closed tetrahedral network depicted in that proof is rotated through space to reverse its orientation, four twists at the vertices are added. Each of these twists contributes a factor of A^n where n, by Lemma 3.10.9, depends on the labels at the vertices. A careful count indicates that these powers of A cancel. □

In [24], Masbaum and Vogel use the recursion relation for the Jones-Wentzl projectors to give a formula for the quantum $6j$-symbols. See also Section 2.8.

3.11.4 Theorem.   *The orthogonality relation and the Elliott-Biedenharn relation hold for the normalized quantum $6j$-coefficients in the following form.*

Orthogonality,

$$\sum_j \Delta_j \Delta_m \begin{bmatrix} b & c & j \\ k & a & n \end{bmatrix}_q \begin{bmatrix} a & b & m \\ c & k & j \end{bmatrix}_q = \delta_{m,n}.$$

Elliott-Biedenharn:

$$\begin{bmatrix} c & d & h \\ g & e & f \end{bmatrix}_q \cdot \begin{bmatrix} b & h & k \\ g & a & e \end{bmatrix}_q =$$

$$\sum_j \Delta_j \begin{bmatrix} b & c & j \\ f & a & e \end{bmatrix}_q \cdot \begin{bmatrix} j & d & k \\ g & a & f \end{bmatrix}_q \cdot \begin{bmatrix} c & d & h \\ k & b & j \end{bmatrix}_q$$

where $\Delta_k = (-1)^{2k}[2k+1]$.

Proof. This is a direct computation. $\square$

3.11.5 Remark. These identities are slightly different than they appear in the classical case. In that case, our choices of the symbol $\begin{bmatrix} b & c & j \\ f & a & e \end{bmatrix}$ were motivated by the desire to make contact with the existing literature, specifically [2]. In the quantum case, these identities coincide exactly with those in [32].

3.12 Theorem.

1.

$$\sum_m (-1)^{a+b+c+k-j-m-n} A^E \begin{Bmatrix} a & b & m \\ c & k & j \end{Bmatrix}_q \begin{Bmatrix} a & c & n \\ k & b & m \end{Bmatrix}_q$$

$$= \begin{Bmatrix} a & c & n \\ b & k & j \end{Bmatrix}_q$$

where

$$E = \{2(a(a+1) + b(b+1) + c(c+1)$$

$$+k(k+1) - j(j+1) - m(m+1) - n(n+1))\}$$

2.

$$\sum_n (-1)^{f+j+n+p} A^{2(f(f+1)+j(j+1)+n(n+1)+p(p+1))}.$$

$$\begin{Bmatrix} a & c & n \\ b & k & j \end{Bmatrix}_q \begin{Bmatrix} a & f & m \\ d & n & c \end{Bmatrix}_q \begin{Bmatrix} b & m & p \\ d & k & n \end{Bmatrix}_q$$

$$= \sum_n (-1)^{k+c+n+m} A^{2(k(k+1)+c(c+1)+n(n+1)+m(m+1))}.$$

$$\left\{ \begin{array}{ccc} b & f & n \\ d & j & c \end{array} \right\}_q \left\{ \begin{array}{ccc} a & n & p \\ d & k & j \end{array} \right\}_q \left\{ \begin{array}{ccc} a & f & m \\ b & p & n \end{array} \right\}_q.$$

3.
$$|2a|2b = \sum_j \Theta(a,b,j) \begin{array}{c} a \setminus \hspace{-0.3em} b \\ j \\ a \hspace{-0.3em} / \hspace{-0.3em} \setminus b \end{array}.$$

4.

$$(-1)^{a+b-k} A^{2(k(k+1)-a(a+1)-b(b+1))} \Theta(a,b,k)$$

$$= \sum_j (-1)^{a+b-j}$$

$$A^{2(a(b+1)+b(b+1)-j(j+1))} \Theta(a,b,j) \left\{ \begin{array}{ccc} a & b & j \\ a & b & k \end{array} \right\}_q.$$

Proof. The proof follows along the same lines as in the proof of Theorem 2.7.14, but crossings are lifted to over and under crossings in a consistent manner so that the diagrams can be isotoped in 3-dimensional space. The remaining details are left to the reader. □

3.12.1 Remark. Similar identities hold for the normalized $6j$-symbols as well.

4 The Quantum Trace and Color Representations

In this section we will define the quantum trace and use it in the case where A is a root of unity to distinguish representations that are not used in the topological applications that we have in mind. The representation theory of $U_q(sl(2))$ in the root of unity case is worked out in [20] under the assumption that E and F are nilpotent and K is of finite order where the orders are determined by the value of q. We will not need to make that assumption as it holds on all the representations that we consider, nor do we need to classify all of the representations of $U_q(sl(2))$.

4.1 The quantum trace. We will define the trace of a linear map $f : W \to W$ in case W is a $U_q(sl(2))$ representation. This notion of trace is quite versatile. In particular, such a trace allowed Jones [12] to define an invariant of knots via braid group representations. Lickorish's [23] combinatorial definition of the Reshetikhin-Turaev invariants [29] is given via the trace. Lickorish's techniques are employed in Section 4.2 in order to prove that the matrix representation of the Temperley-Lieb algebra is faithful for generic values of q.

Finally, we will use the trace to characterize certain representations in the case when $A = e^{i\pi/(2r)}$ (or any other primitive 4rth root of unity). The tensor power $(V_A^{1/2})^{\otimes n}$ decomposes as a direct sum of a trace 0 summand and a sum of representations V_A^j where j is taken from the finite set $\{0, 1/2, \ldots, (r-2)/2\}$. By excluding the representations that have trace 0, we will be able to define the Turaev-Viro [32] invariant as a state summation.

4.1.1 Definition. Let W be a representation space for $U_q(sl(2))$. Then the *quantum trace* of a linear map $f : W \to W$ is the ordinary trace of the map $K^2 \circ f$.

In particular, if

$$L : (V_A^{1/2})^{\otimes n} \to (V_A^{1/2})^{\otimes n}$$

denotes a linear map, then

$$tr_q(L) = tr(K^2 \circ L).$$

Consider the bases $\{f_S : S \subset \{1, 2, \ldots, n\}\}$ and $\{\bar{f}_S : S \subset \{1, 2, \ldots, n\}\}$ for $(V_A^{1/2})^{\otimes n}$ where $f_s = x_1^S \otimes \cdots \otimes x_n^S$ and $\bar{f}_s = \bar{x}_n^S \otimes \cdots \otimes \bar{x}_1^S$ with

$$x_j^S = \begin{cases} x & \text{if } j \notin S \\ y & \text{if } j \in S \end{cases}$$

and

$$\bar{x}_j^S = \begin{cases} x & \text{if } j \in S \\ y & \text{if } j \notin S \end{cases}.$$

In the first basis the matrix representation of L is

$$(L_{RS})_{R,\, S \,\subset\, \{1, \ldots, n\}}$$

where

$$L(f_S) = \sum_R L_{RS} f_R$$

and the sum is taken over all subsets $R \subset \{1, \ldots, n\}$.

4.1.2 Lemma.

$$\overset{n}{\cap} \circ (L \otimes |_n) \circ \overset{n}{\cup} = (-1)^n tr_q(L)$$

$$= \sum_S q^{n-2|S|} L_{SS}$$

where the sum is taken over the set of all subsets $S \subset \{1, \ldots, n\}$.

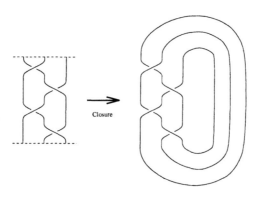

Figure 7: The braid closure of a tangle

Proof. The second equality follows because $K f_S = A^{2(r/2-s/2)} f_S = A^{n-2|S|} f_S$ where $r = \#\{k : x_k^S = x\}$ and $s = \#\{k : x_k^S = y\}$. For the first formula, we compute using Lemma 3.6.4

$$\overset{n}{\cap} \circ (L \otimes |_n) \circ \overset{n}{\cup} (1) = \overset{n}{\cap} \circ (L \otimes |_n)$$
$$\left(i^n \sum_{S \subset \{1,\dots,n\}} (-1)^{|S|} A^{n-2|S|} f_S \otimes \bar{f}_S \right)$$
$$= i^n \overset{n}{\cap} \left(\sum_{S \subset \{1,\dots,n\}} (-1)^{|S|} A^{n-2|S|} L(f_S) \otimes \bar{f}_S \right)$$
$$= i^n \sum_{R,S \subset \{1,\dots,n\}} (-1)^{|S|} A^{n-2|S|} L_{RS} \overset{n}{\cap} (f_R \otimes \bar{f}_S)$$
$$= i^n \sum_{S \subset \{1,\dots,n\}} (-1)^{|S|} A^{n-2|S|} (iA)^{n-2|S|} L_{SS}$$
$$= (-1)^n \sum_{S \subset \{1,\dots,n\}} A^{2(n-2|S|)} L_{SS}$$

This completes the proof. $\square$

4.1.3 Remark. The point of the Lemma is, for maps L defined on $(V_A^{1/2})^{\otimes n}$ in terms of Temperley-Lieb elements, the trace of the operator can be computed by the bracket expansion of the braid closure of the operator; the *braid closure* of a tangle diagram is depicted in Figure 7. In particular, we apply the Lemma to the Jones-Wentzl projectors to obtain the following:

4.1.4 Proposition. *Suppose that $A^{4r} \neq 1$, or if A is a primitive $4r$th root of unity, then suppose that $n \leq r - 1$.*

$$tr_q(\overset{A}{\underset{n}{+}}) = [n + 1].$$

In particular, if A is a primitive $4r$th root of unity and if $n = r - 1$, then

$$tr_q(\overset{A}{\underset{n}{+}}) = 0.$$

Proof. The proof follows from comments made in Remark 3.7.2.
□

4.1.5 Definition. Now suppose $A = e^{\pi i/(2r)}$, and as usual, let $q = A^2$. The $U_q(sl(2))$ representation W is said to be *of (quantum) trace 0* if $tr_q(f) = tr(K^2 \circ f) = 0$ for every $U_q(sl(2))$ invariant map $f : W \to W$.

4.2 A bilinear form on tangle diagrams. Recall that a tangle diagram is a diagram of arcs embedded in a thin rectangular prism obtained by projecting in the short direction onto a rectangular face and depicting crossing information by breaking the arc at an intersection point that is farthest away from the plane of projection. Now suppose that such a diagram has the same number of incoming and outgoing strands, say n. Such a tangle diagram (in general position with respect to a height function) represents a map $(V_A^{1/2})^{\otimes n} \to (V_A^{1/2})^{\otimes n}$ via the association of ∩, and ∪ to the maximal and minimal points respectively, and the association of the bracket identity to each crossing. Regularly isotopic diagrams where the end points are fixed during the isotopy determine the same maps.

Let T_n denote the set linear combinations of such tangle diagrams. The bilinear form

$$\langle \cdot, \cdot \rangle : T_n \times T_n \to \mathbf{C}$$

is defined as the linear extension of the map defined on tangles S and M by:

$$\langle S, M \rangle = \overset{n}{\cap} \circ (S \otimes M) \circ \overset{n}{\cup}$$

$$= \overset{n}{\cap} \circ ((S \circ W) \otimes |_n) \circ \overset{n}{\cup}$$

where W is a rotation of the tangle M through $180°$ about an axis perpendicular to the plane of the diagram. The second equality follows because the diagrams representing these tangles are regularly isotopic. The image of $\langle \cdot, \cdot \rangle$ is $\mathbf{C}$ by evaluating the diagrams via the bracket identity; the parameter A is assumed to be a nonzero complex number.

The Temperley-Lieb algebra, TL_n, has dimension $\binom{2n}{n}/(n+1)$. A vector space basis is a certain set of monomials in the generators $\{1, h_1, \cdots, h_{n-1}\}$. Because of the relations among the products of these generators, any monomial can be put into a normal form. We will not list the normal form monomials here, but we will assume that these form a basis.

We can mod out the free module generated by regular isotopy classes of tangle diagrams by the ideal generated by the bracket relation and the relation that the loop value is δ. In this way, every n-string tangle diagram can be written as a linear combination of elements in the basis of TL_n. A matrix representation of the inner product $\langle \cdot, \cdot \rangle$ is given as the collection

$$\{\langle w_j(h_1, \cdots, h_{n-1}), w_k(h_1, \cdots, h_{n-1})\rangle\}$$

where w_j and w_k range over the set of normal form monomials.

For example, when $n = 3$, the matrix is the matrix

$$\begin{bmatrix} \delta^3 & \delta^2 & \delta^2 & \delta & \delta \\ \delta^2 & \delta & \delta^3 & \delta^2 & \delta^2 \\ \delta^2 & \delta^3 & \delta & \delta^2 & \delta^2 \\ \delta & \delta^2 & \delta^2 & \delta & \delta^3 \\ \delta & \delta^2 & \delta^2 & \delta^3 & \delta^3 \end{bmatrix}$$

where δ is the value associated to any simple closed curve, and the ordered basis of the algebra TL_3 is $(|_3, h_1, h_2, h_1h_2, h_2h_1)$.

Throughout this paper we have assumed that $\delta = -A^2 - A^{-2}$ because this is the loop value associated to the identification

$$\text{\small⊔⊓} \mapsto [0, iA, -iA^{-1}, 0]^t \cdot [0, iA, -iA^{-1}, 0].$$

More generally, δ is the loop value in $\mathbf{C}$ of the Temperley-Lieb algebra regardless of the representation.

4.2.1 Proof of Theorem 3.3.4. This proof was indicated to us by Paul Melvin. Let T_n denote the matrix of the inner product $\langle \cdot, \cdot \rangle$ defined on n-string tangles with respect to the basis of normal form monomials in the generators $1, h_1, \ldots, h_{n-1}$. Observe that $\det T_n$ is a polynomial function of the loop value δ. Under the given representation $\delta = -A^2 - A^{-2}$.

The representation,

$$\text{\small⊔⊓} \mapsto [0, iA, -iA^{-1}, 0]^t \cdot [0, iA, -iA^{-1}, 0],$$

of the Temperley-Lieb algebra into a matrix algebra is faithful when the matrix T_n is non-singular.

In [21] Ko and Smolinski show that all the roots of the equation $\det T_n = 0$ are of the form $\delta = 2\cos k\pi/(m+1)$ where $1 \leq m \leq n$ and $1 \leq k \leq m$. So T_n is non-singular unless A is a $4r$th root of unity and $n \geq r-1$. This completes the proof. $\square$

4.2.2 Remark. The singularity of the pairing $\langle \cdot, \cdot \rangle$ when $n \geq r-1$ is a key ingredient in Lickorish's [23] construction of the Reshetikhin-Turaev invariants. Ko and Smolinsky give an elementary combinatorial proof of this singularity via a recursive construction of the Temperley-Lieb basis. The proof of singularity is originally due to Jones [13].

Our proof of faithfulness breaks down in case the matrix T_n is singular, but according to [7] the representation θ_A is faithful in all cases except possibly when $A = -1$. Their proof follows along the same lines as ours in the classical case which also covers the case of $A = -1$.

4.3 Color representations. We keep to the case that A is a primitive $4r$th root of unity with the integer $r \geq 3$. In this case $[r] = 0$. For such values of A define the set of colors to be the set $\{0, 1/2, 1, 3/2, \ldots, (r-2)/2\}$. We will consider representations V_A^j where j is a color, for these representations are irreducible and by computing modulo the trace zero representations, we will still have a Clebsch-Gordan theory.

4.3.1 Lemma. *The representation V_A^a is not of trace 0 if a is a color. It is of trace 0 if $a = (r-1)/2$.*

Proof. A computation with weight vectors shows that the representation V_A^a is irreducible when $a \leq (r-1)/2$. By Schur's Lemma, any $U_q(sl(2))$ map is a multiple of the identity. It suffices to compute $tr_q(\mapsto_{2a}^A) = [2a+1]$ (Lemma 4.1.4); this trace is non-zero unless $a = (r-1)/2$. $\square$

4.3.2 Definition. A *q-admissible triple* is a triple (a, b, j) of colors such that

1. $a + b + j$ is an integer;

2. $a + b - j$, $b + j - a$, and $j + a - b$ are all ≥ 0;

3. $a + b + j \leq r - 2$.

The role of the last condition is as follows. When A is a primitive 4rth root of unity and when (a, b, j) is q-admissible, then the map $\bigvee_{j}^{ab}$ is non-zero (Theorem 3.6.6). Consider the computation of the closed Θ-network, $\Theta(a, b, j)$, given in Theorem 3.7.1. In the root of unity case, $\Theta(a, b, j) = 0$ when $a + b + j > r - 2$. The interpretation is that the composition $\bigwedge_{ab}^{j} \circ \bigvee_{j}^{ab}$ vanishes because $\bigvee_{j}^{ab}$ maps into a summand of $(V^{1/2})^{\otimes 2(a+b)}$ that is of trace zero.

4.3.3 Lemma. *For colors b,*

$$V_A^b \otimes V_A^{1/2} \simeq \begin{cases} V_A^{1/2} & \text{if } b = 0; \\ V_A^{b+1/2} \oplus V_A^{b-1/2} & \text{if } 1/2 \leq b \leq (r-2)/2. \end{cases}$$

Observe that $V_A^{b+1/2}$ has trace zero if $b = (r-2)/2$.

Proof. The relevant maps $\bigvee_{j}^{b,1/2}$ are defined and non-zero for $j = b \pm 1/2$; so the result follows by a dimension count. $\square$

4.3.4 Lemma. *If M is a finite dimensional $U_q(sl(2))$ representation of trace zero, then $M \otimes V_A^{1/2}$ also has trace zero.*

Proof. Let $\phi : M \otimes V_A^{1/2} \to M \otimes V_A^{1/2}$ be a $U_q(sl(2))$ invariant map. Define an invariant map $\tilde{\phi} : M \to M$ by the formula

$$\tilde{\phi} = (1_M \otimes \cap) \circ (\phi \otimes |) \circ (1_M \otimes \cup)$$

mapping

$$M = M \otimes \mathbf{C} \to M \otimes V_A^{1/2} \otimes V_A^{1/2} \to M \otimes V_A^{1/2} \otimes V_A^{1/2} \to M \otimes \mathbf{C} = M.$$

An algebraic computation shows that $tr_q(\phi) = -tr_q(\tilde{\phi})$. Since M has trace zero, $tr_q(\tilde{\phi}) = 0$, and so $tr_q(\phi) = 0$, which completes the proof. $\square$

4.3.5 Lemma. *For colors a and b, there is a subrepresentation $U_{a,b}$ of $V_A^a \otimes V_A^b$ such that*

1. $U_{a,b}$ has trace zero, and

2. $V_A^a \otimes V_A^b \simeq U_{a,b} \oplus \left(\bigoplus_j V_A^j \right)$ where the sum is over all colors j such that (a, b, j) is a q-admissible triple.

Proof. The proof is by induction on b. The case $b = 0$ is trivial. The case $b = 1/2$ is Lemma 4.3.3.

Now suppose that $1 \leq b \leq (r - 2)/2$. On the one hand, the inductive hypothesis applied to $b - 1$ shows that

$$V_A^a \otimes V_A^{b-1/2} \otimes V_A^{1/2} = V_A^a \otimes (V_A^{b-1/2} \otimes V_A^{1/2})$$

$$\simeq V_A^a \otimes (V_A^b \oplus V_A^{b-1})$$

$$\simeq V_A^a \otimes V_A^b \oplus \left(\bigoplus_{j:(a,b-1,j)q-\text{admissible}} V_A^j \right) \oplus U_{a,b-1}$$

where $U_{a,b-1}$ has trace zero.

On the other hand, Lemma 4.3.4 and the inductive hypothesis applied to $b - 1/2$ show that

$$V_A^a \otimes V_A^{b-1/2} \otimes V_A^{1/2} = (V_A^a \otimes V_A^{b-1/2}) \otimes V_A^{1/2}$$

$$\simeq \left(\left(\bigoplus_{k:(a,b-1/2,k)q-\text{admissible}} V_A^k \right) \oplus U_{a,b-1/2} \right) \otimes V_A^{1/2}$$

$$\simeq \bigoplus_{(k,j)} V_A^j \oplus U$$

where the sum is over the pairs (k, j) such that both $(a, b - 1/2, k)$ and $(k, 1/2, j)$ are q-admissible, and where the representation U has trace zero.

Comparing the two expressions and using the Remak-Krull-Schmidt Theorem [9], we find that to finish the proof we must show that

$$\bigoplus_{(k,j)} V_A^j \simeq \left(\bigoplus_{j:(a,b,j)q-\text{admissible}} V_A^j \right) \oplus \left(\bigoplus_{j:(a,b-1,j)q-\text{admissible}} V_A^j \right)$$

where the first sum is over the pairs (k, j) such that both $(a, b - 1/2, k)$ and $(k, 1/2, j)$ are q-admissible. This is a purely combinatorial question. One checks tediously but straight-forwardly that for fixed $1 \leq b \leq (r - 2)/2$ and $0 \leq a \leq (r - 2)/2$, the following map defines a bijection between $X = \{(k, j)\}$ and $Y = \{(a, b, j)\} \cup \{(a, b - 1, j)\}$.

$$(k, j) \mapsto \begin{cases} (a, b, j) & \text{if } j = a - b \text{ or } (j > k \text{ and } j \neq r - 1 - a - b); \\ (a, b - 1, j) & \text{if } j = r - 1 - a - b \text{ or } (j < k \text{ and } j \neq a - b). \end{cases}$$

This completes the proof. $\square$

4.3.6 Lemma. *If a is a color and M is a finite dimensional $sl_q(2)$ representation of trace zero, then $M \otimes V_A^a$ also has trace zero.*

Proof. The proof is by induction on a. The case $a = 0$ is trivial, and the case $a = 1/2$ is just Lemma 4.3.4. Now suppose that $1 \le a \le (r-2)/2$. By Lemma 4.3.3, V_A^a is a summand of $V_A^{a-1/2} \otimes V_A^{1/2}$, and it follows immediately that $M \otimes V_A^a$ is a summand of $M \otimes (V_A^{a-1/2} \otimes V_A^{1/2})$. On the other hand, Lemma 4.3.4 and the inductive hypothesis applied to $a - 1/2$ show that $(M \otimes V_A^{a-1/2}) \otimes V_A^{1/2}$ is a representation of trace 0. Since every summand of a trace 0 representation is of trace 0, the representation $M \otimes V_A^a$ is of trace 0. $\square$

4.3.7 Proposition. *Let A denote a primitive $4r$th root of unity. Let a_ℓ be in the set of colors for $\ell = 1, \dots, N$. Then*

$$V_A^{a_1} \otimes \cdots \otimes V_A^{a_N} \simeq \left(\bigoplus_j V_A^j \right) \oplus U$$

where the first sum is over an appropriate set of colors with possible multiplicity and U is of trace 0.

Proof. We induct on the number of factors N. For $N = 1$ the result is trivial. For $N = 2$, Lemma 4.3.5 applies. For $N > 2$, we have by the inductive hypothesis:

$$V_A^{a_1} \otimes \cdots \otimes V_A^{a_{N-1}} \otimes V_A^{a_N} \simeq \left(\bigoplus_j V_A^j \oplus U \right) \otimes V_A^{a_N}$$

$$= \bigoplus_j \left(V_A^j \otimes V_A^{a_N} \right) \oplus (U \otimes V_A^{a_N})$$

The first term decomposes as a direct sum of colors and a trace 0 piece by the case $N = 2$. The second term is of trace 0 by Lemma 4.3.6. This completes the proof. $\square$

4.3.8 Proposition. *For colors a and b,*

$$V_A^a \otimes V_A^b = \left(\bigoplus_j (\mu_a \otimes \mu_b) \circ \bigvee_j^{ab} \circ \phi_j(V_A^j) \right) \oplus U'$$

where the sum is over all colors j such that (a, b, j) is a q-admissible triple, and the summand U' is a subrepresentation of trace 0.

Proof. First we observe that in the case of colors k the Jones-Wentzl projectors $+^A_{2k}$ are defined. Thus $\bigvee^{ab}_{j}$ is defined, for (a, b, j) a q-admissible triple, and the map $\phi_j : V^j_A \to (V^{1/2}_A)^{\otimes 2j}$ is defined.

Let $V^a_A \otimes V^b_A = W' \oplus U'$ be a direct sum decomposition where W' is a direct sum of color representations and U' is a subrepresentation of trace 0. By Lemma 4.3.5, we just have to prove that under the stated conditions $(\mu_a \otimes \mu_b)(\bigvee^{ab}_{j}(\phi_j(V^j_A)))$ is not contained in U'.

Let $U'' = \phi_a \otimes \phi_b(U') \subset (V^{1/2}_A)^{\otimes 2a+2b}$. Since $\phi_a \otimes \phi_b(V^a_A \otimes V^b_A)$ is a direct summand of $(V^{1/2}_A)^{\otimes 2(a+b)}$ and U' is a a direct summand of $V^a_A \otimes V^b_A$, U'' is a direct summand of $(V^{1/2}_A)^{\otimes 2(a+b)}$. We recall that for any color ℓ, the multiplication map $\mu_\ell|_{\phi_\ell(V^\ell_A)}$ is an isomorphism inverse to ϕ_ℓ. So $(\mu_a \otimes \mu_b)|_{\phi_a \otimes \phi_b(V^a_A \otimes V^b_A)}$ is an isomorphism that is the inverse of $\phi_a \otimes \phi_b$. Now the image of $\bigvee^{ab}_{j}$ is the subspace $\bigvee^{ab}_{j}\phi_j(V^j_A)$ of $\phi_a \otimes \phi_b(V^a_A \otimes V^b_A)$. If we show that the subspace $\bigvee^{ab}_{j}\phi_j(V^j_A)$ intersects the trace 0 summand U'' trivially, it will follow that $\left(\mu_a \otimes \mu_b \circ \bigvee^{ab}_{j} \circ \phi_j(V^j_A)\right) \cap U' = (0)$.

Thus we complete the proof by showing that $\bigvee^{ab}_{j}(\phi_j(V^{1/2}_A)^{\otimes 2j})$ is not contained in any trace 0 summand, U, of $(V^{1/2}_A)^{\otimes 2(a+b)}$. Suppose it were. Consider the composition $f = \bigvee^{ab}_{j} \circ \bigwedge^{j}_{ab}$. This is a $U_q(sl(2))$ endomorphism of $(V_A^{1/2})^{\otimes 2(a+b)}$. Its quantum trace is the value $\Theta(a, b, j) \neq 0$ (See Remark 3.7.4). Let $W \oplus U$ denote a direct sum decomposition of $(V_A^{1/2})^{\otimes 2(a+b)}$ where W is a subrepresentation complementary to U, and let p_W and p_U denote the projections onto the indicated summands. Now we compute

$$
\begin{aligned}
tr_q(f) \;=\;& tr(K^2 \circ f) \\
=\;& tr(K^2 \circ p_W \circ f \circ p_W) + tr(K^2 \circ p_U \circ f \circ p_W) \\
+\;& tr(K^2 \circ p_W \circ f \circ p_U) + tr(K^2 \circ p_U \circ f \circ p_U) \\
=\;& 0
\end{aligned}
$$

The first term is 0 because the image of f is assumed to be in U; the next two terms are 0 because in a block matrix representation of these maps, the two diagonal blocks are 0. The last term is 0 because it equals the quantum trace of the $U_q(sl(2))$ invariant map $p_U \circ f|_U$, and U has trace zero. This contradiction completes the proof. $\square$

4.4 The quantum $6j$-symbol — root of unity case.

In order to define the $6j$-symbol in the case that A is a primitive 4rth root of unity, we will need to decompose the tensor product of three color representations into a direct sum of colors and a summand that is of trace 0. We will mod out by the trace 0 piece and construct bases for invariant maps into the quotient. The $6j$-symbol, then, will be defined as a change of basis matrix in a family of maps from a color representation to the quotient of the tensor product by a maximal trace 0 summand. Let us turn to the construction.

4.4.1 Proposition.   *Let A denote a primitive 4rth root of unity, and let $a, b, c \in \{0, 1/2, \ldots, (r-2)/2\}$ — the set of colors.*

Let $\mu_{abc} = \mu_a \otimes \mu_b \otimes \mu_c$. Then

$$V_A^a \otimes V_A^b \otimes V_A^c =$$

$$= \bigoplus_{\{n,k: \ (a,b,n)\&(c,n,k) \ \ q\text{-admis.}\}} \mu_{abc}\circ \qquad n \qquad \circ\phi_k(V_A^k) \oplus U_1$$

$$= \bigoplus_{\{j,k: \ (b,c,j)\&(a,j,k) \ \ q\text{-admis.}\}} \mu_{abc}\circ \qquad j \qquad \circ\phi_k(V_A^k) \oplus U_2$$

where U_1 and U_2 are trace 0 subrepresentations of $V_A^a \otimes V_A^b \otimes V_A^c$.

Proof. We have isomorphisms $(\star)$

$$(V_A^a \otimes V_A^b) \otimes V_A^c \simeq \bigoplus_{\{n:(a,b,n) \ \text{is} \ q\text{-ad.}\}} V_A^n \otimes V_A^c \oplus (U_{a,b} \otimes V_A^c)$$

$$\simeq \bigoplus_{\{n:(a,b,n) \ \text{is} \ q\text{-ad.}\}} \left(\bigoplus_{\{k:(n,c,k) \ \text{is} \ q\text{-ad.}\}} V_A^k \oplus U_{n,c} \right) \oplus (U_{a,b} \otimes V_A^c)$$

$$= \bigoplus_{\{n,k:(a,b,n)\&(n,c,k) \ \text{are} \ q\text{-admis.}\}} V_A^k$$

$$\oplus \left(\bigoplus_{\{n:(a,b,n) \ \text{is} \ q\text{-ad.}\}} U_{n,c} \oplus (U_{a,b} \otimes V_A^c) \right)$$

where the term $\bigoplus_{\{n:(a,b,n) \ \text{is} \ q\text{-admis.}\}} U_{n,c} \oplus (U_{a,b} \otimes V_A^c)$ on the right of the last equality is of trace 0. Furthermore,

$$(V_A^a \otimes V_A^b) \otimes V_A^c$$

$$= \bigoplus_{n:(a,b,n) \text{ is } q\text{-ad.}} \left(\mu_{ab} \circ \bigvee_n^{\,ab} \circ \phi_n(V_A^n) \right) \otimes V_A^c \oplus (U \otimes V_A^c)$$

$$= \bigoplus_{n:(a,b,n) \text{ is } q\text{-ad.}} \mu_{abc} \circ \left(\bigvee_n^{\,ab} \otimes \Big|_c \right) \circ (\phi_n \otimes \phi_c)(V_A^n \otimes V_A^c) \oplus (U \otimes V_A^c)$$

$$= \bigoplus_{n:(a,b,n) \text{ is } q\text{-ad.}} \left(\bigoplus_{k:(n,c,k) \text{ is } q\text{-ad.}} \mu_{abc} \circ \left(\bigvee_n^{\,ab} \otimes \Big|_c \right) \circ \bigvee_k^{\,nc} \circ \phi_k(V_A^k) \right.$$

$$\left. \oplus \mu_{abc} \circ \left(\bigvee_n^{\,ab} \otimes \Big|_c \right) \circ (\phi_n \otimes \phi_c)(U') \right) \oplus (U \otimes V_A^c)$$

$$= \left(\bigoplus_{n,k:(a,b,n)\&(n,c,k) \text{ are } q\text{-ad.}} \mu_{abc} \circ \quad \begin{matrix} a \diagdown \quad b \quad c \\ \quad n \diagup \\ \quad \diagdown \\ \quad k \end{matrix} \quad \circ \phi_k(V_A^k) \right)$$

$$\bigoplus U_1$$

By computation $(\star)$ and the Remak-Krull-Schmidt Theorem, the subrepresentation U_1 must be of trace 0. This completes the proof of the first equality.

Now we turn to establish the second equality. We claim that up to isomorphism,

$$V_A^a \otimes V_A^b \otimes V_A^c \simeq \left(\bigoplus_{\{j,k:(b,c,j)\&(a,j,k) \text{are } q \text{ admis.}\}} V_A^k \right) \oplus U'$$

where U' has trace 0. By the computation $(\star)$ and the Remak-Krull-Schmidt Theorem, it suffices to show $(\star\star)$

$$\bigoplus_{\{j,k:(b,c,j)\&(a,j,k) \text{are } q \text{ admis.}\}} V_A^k \simeq \bigoplus_{\{n,k:(a,b,n)\&(n,c,k) \text{are } q \text{ admis.}\}} V_A^k$$

The proof of ($\star\star$) reduces to a tedious combinatorial exercise, namely to show that for each color k there is a bijection between the sets

$$\{j \in \{0, 1, \ldots, (r-2)/2\} : (b, c, j) \ \& \ (a, j, k) \text{ are } q \text{ admissible}\}$$

and

$$\{n \in \{0, 1, \ldots, (r-2)/2\} : (a, b, n) \ \& \ (n, c, k) \text{ are } q \text{ admissible}\}.$$

Finally, we compute

$$V_A^a \otimes (V_A^b \otimes V_A^c)$$

$$= (V_A^a \otimes U) \oplus \bigoplus_{j:(b,c,j) \text{ is } q\text{-admis.}} V_A^a \otimes \left(\mu_{bc} \circ \bigvee\limits_{j}^{bc} \circ \phi_j(V_A^j) \right)$$

$$= (V_A^a \otimes U) \oplus \bigoplus_{j:(b,c,j) \text{ is } q\text{-ad.}} \mu_{abc} \circ \left(\Big|_a \otimes \bigvee\limits_{j}^{bc} \right) \circ (\phi_a \otimes \phi_j)(V_A^a \otimes V_A^j)$$

$$= (V_A^a \otimes U)$$

$$\bigoplus_{j:(b,c,j) \text{ is } q\text{-ad.}} \left(\bigoplus_{k:(a,j,k) \text{ is } q\text{-ad.}} \mu_{abc} \circ \left(\Big|_a \otimes \bigvee\limits_{j}^{bc} \right) \circ \bigvee\limits_{k}^{aj} \circ \phi_k(V_A^k) \right.$$

$$\left. \oplus \mu_{abc} \circ \left(\Big|_a \otimes \bigvee\limits_{j}^{bc} \right) \circ (\phi_a \otimes \phi_j)(U') \right)$$

$$= \left(U_2 \oplus \bigoplus_{j,k:(b,c,j)\&(a,j,k) \text{ are } q\text{-ad.}} \mu_{abc} \circ \ \substack{a \diagdown \ b \diagup c \\ \diagdown \Y \diagup \ j \\ k |} \circ \phi_k(V_A^k) \right)$$

By ($\star\star$) and Remak-Krull-Schmidt, the subrepresentation U_2 must also be of trace 0. This completes the proof. $\square$

4.4.2 Remark. The splittings of Proposition 4.4.1 are not unique because the maximal trace zero summands of $V_A^a \otimes V_A^b \otimes V_A^c$ are not unique.

4.4.3 Lemma. *Let A be a 4rth root of unity, and let a, b, and c be colors. Then all maximal trace zero summands of $V_A^a \otimes V_A^b \otimes V_A^c$ contain precisely the same irreducible $U_q(sl(2))$ submodules.*

Proof. Let U_1 and U_2 be maximal trace zero summands, say

$$V_A^a \otimes V_A^b \otimes V_A^c = W \oplus U_1,$$

where, by Proposition 4.3.7, W is completely reducible. Let V be an irreducible submodule of $V_A^a \otimes V_A^b \otimes V_A^c$ such that $V \not\subset U_1$. By irreducibility of V, $V \cap U_1 = (0)$, and by complete reducibility of W, $V \oplus U_1$ is a summand of $V_A^a \otimes V_A^b \otimes V_A^c$. Moreover, V is a color representation.

Suppose that $V \subset U_2$. Let p_V be a $U_q(sl(2))$ invariant projection from $V_A^a \otimes V_A^b \otimes V_A^c$ to V and consider its restriction $\mathrm{res}_{U_2}(p_V)$ to U_2. On the one hand, $tr_q(\mathrm{res}_{U_2}(p_V)) = 0$, because U_2 is a representation of trace zero. On the other hand, $tr_q(\mathrm{res}_{U_2}(p_V)) = tr_q(1_V) \neq 0$, because V is a color representation. This contradiction proves that $V \not\subset U_2$. The proof of the Lemma is complete. □

4.4.4 Lemma. *Let A be a primitive 4rth root of unity, let a, b, c, and k be colors, and let*

$$T : V_A^k \to V_A^a \otimes V_A^b \otimes V_A^c$$

be a $U_q(sl(2))$ invariant map. Then there are unique complex numbers d_n and map $S : V_A^k \to V_A^a \otimes V_A^b \otimes V_A^c$ such that

$$T = \sum_{\substack{\{n : (a,b,n) \& (n,c,k) \\ \text{are } q\text{-admis.}\}}} d_n (\mu_a \otimes \mu_b \otimes \mu_c) \circ \quad \text{(diagram)} \quad \circ \phi_k + S$$

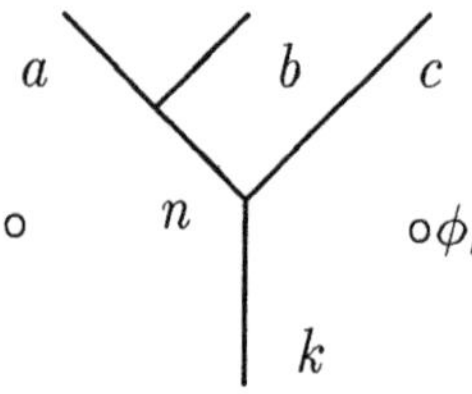

where the diagram shows strands labeled a, b, c at top, meeting at node n, with strand k at bottom.

and $S(V_A^k)$ is contained in a trace zero summand of $V_A^a \otimes V_A^b \otimes V_A^c$.

Proof. The existence of the numbers d_n and map S is immediate from Proposition 4.4.1. Suppose that d_n' and S' were another solution, so that

$$\sum_{\{n:(a,b,n)\&(n,c,k) \text{ are } q\text{-admis.}\}} (d_n - d_n') (\mu_a \otimes \mu_b \otimes \mu_c) \circ$$

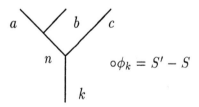

$$\circ \phi_k = S' - S$$

By the preceding lemma, the image of the right side $S' - S$ lies within any maximal trace zero summand of $V_A^a \otimes V_A^b \otimes V_A^c$, and by Proposition 4.4.1 the image of the left side must intersect such a summand trivially. Therefore both sides are the zero maps. Thus $S = S'$, and by the linear independence of the maps

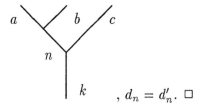

$, d_n = d'_n.$ $\square$

4.4.5 Definition.   In light of the preceding Lemma, we can define in the case that A is a primitive $4r$th root of unity the *6j-symbol* to be the coefficient $\left\{ \begin{matrix} a & b & n \\ c & k & j \end{matrix} \right\}_q$ in the following equation:

$$\mu_{abc} \circ \ \overset{a \quad b \qquad c}{\underset{k}{\bigvee}}_{j} \ \circ \phi_k$$

$$= \sum_n \left\{ \begin{matrix} a & b & n \\ c & k & j \end{matrix} \right\}_q \mu_{abc} \circ \ \overset{a \quad b \quad c}{\underset{k}{\bigvee}_n} \ \circ \phi_k + S$$

where S maps into a summand of trace 0, $\mu_{abc} = \mu_a \otimes \mu_b \otimes \mu_c$, and the sum is taken over the set of colors n such that the triples (a, b, n) and (n, c, k) are q-admissible.

4.4.6 Theorem.   *Let A denote a primitive $4r$th root of unity. Then the quantum $6j$-symbols satisfy the Elliott-Biedenharn and orthogonality identities as stated in Theorems 3.8.6 and 3.8.5 where the sums are over q-admissible indices. Moreover for such values of A, the normalized $6j$-symbols are defined and the identities stated in Theorem 3.11.4 hold.*

Proof. First we will show that

$$
\raisebox{-1em}{tree}_{a\ b\ c\ j\ k} = \sum_n \left\{ \begin{matrix} a & b & n \\ c & k & j \end{matrix} \right\}_q \raisebox{-1em}{tree}_{a\ b\ c\ n\ k}
$$

where by equality we mean the following: The value of any closed network in which the tree on the left appears is equal to the weighted sum of the values of the closed networks obtained by replacing the given tree by the trees on the right; the weights are the indicated $6j$-symbols. We have

$$
\raisebox{-1em}{tree}_{a\ b\ c\ j\ k} = \phi_{abc} \circ \mu_{abc} \circ \raisebox{-1em}{tree}_{a\ b\ c\ j\ k} \circ \phi_k \circ \mu_k
$$

$$
= \sum_n \left\{ \begin{matrix} a & b & n \\ c & k & j \end{matrix} \right\}_q
$$

$$
\times \phi_{abc} \circ \mu_{abc} \circ \raisebox{-1em}{tree}_{a\ b\ c\ n\ k} \circ \phi_k \circ \mu_k + \phi_{abc} \circ S \circ \mu_k
$$

$$= \sum_n \left\{ \begin{matrix} a & b & n \\ c & k & j \end{matrix} \right\}_q \quad \text{(diagram)} \quad + \quad \phi_{abc} \circ S \circ \mu_k$$

When the map $\phi_{abc} \circ S \circ \mu_k$ is inserted into a closed network, the network can be rearranged in its planar representation so that its value is $(-1)^k tr_q(\phi_{abc} \circ S \circ \mu_k \circ T)$ where T is some $U_q(sl(2))$ invariant map. The image of S is in a summand of trace 0; thus this value is 0.

We employ the diagrammatic techniques given in the proof of Lemma 2.6.6 to show that

$$\text{(diagram)} = \sum_u \left\{ \begin{matrix} m & p & u \\ t & s & r \end{matrix} \right\} \text{(diagram)}$$

in the sense of inserting these diagrams into any closed network. As a consequence, we have

$$\text{(diagram)} = \sum_j \left\{ \begin{matrix} b & c & j \\ k & a & n \end{matrix} \right\}_q \text{(diagram)}$$

where the sum is over all j such that the appropriate triples are q-admissible and the equality is in the sense of inserting the diagrams into closed networks.

We form a closed network by reflecting 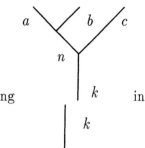 in

the horizontal axis to obtain the diagram

This is juxtaposed (with the upper index n changed to s) to

the top of 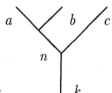 to form the network:

$$N = N(a, b, c, n, j, k) =$$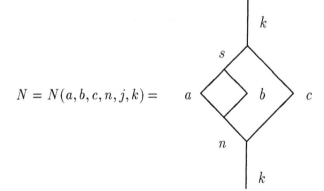

Then the network, N, is closed by joining the arcs that are labeled by k. Let $\hat{N}$ denote this closed network, or its value in **C**. We obtain:

$$\delta_{s,n} \Theta(a, b, s) \Theta(s, c, k) / \Delta_s =$$

$$\hat{N} = \sum_m \sum_j \left\{ \begin{matrix} b & c & j \\ k & a & n \end{matrix} \right\}_q \left\{ \begin{matrix} a & b & m \\ c & k & j \end{matrix} \right\}_q \hat{N}$$

$$= \sum_m \sum_j \left\{ \begin{matrix} b & c & j \\ k & a & n \end{matrix} \right\}_q \left\{ \begin{matrix} a & b & m \\ c & k & j \end{matrix} \right\}_q \delta_{s,m} \Theta(a,b,s)\Theta(s,c,k)/\Delta_s$$

from which it follows that

$$\delta_{m,n} = \sum_j \left\{ \begin{matrix} b & c & j \\ k & a & n \end{matrix} \right\}_q \left\{ \begin{matrix} a & b & m \\ c & k & j \end{matrix} \right\}_q .$$

Now we establish the Elliott-Biedenharn identity using the same technique. As in the proof of Theorem 2.6.7, we have the identity —

$$\sum_{h,k} \left\{ \begin{matrix} c & d & h \\ g & e & f \end{matrix} \right\}_q \cdot \left\{ \begin{matrix} b & h & k \\ g & a & e \end{matrix} \right\}_q T_{h,k}$$

$$= \sum_{h,k,j} \left\{ \begin{matrix} b & c & j \\ f & a & e \end{matrix} \right\}_q \cdot \left\{ \begin{matrix} j & d & k \\ g & a & f \end{matrix} \right\}_q \cdot \left\{ \begin{matrix} c & d & h \\ k & b & j \end{matrix} \right\}_q T_{h,k}$$

where T is the tree depicted on the left below — that holds in the sense of insertion into closed networks.

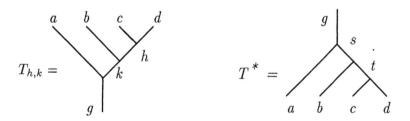

Let $T^* = T^*(a,b,c,d,s,t,g)$ denote the mirror image of T through a horizontal axis as shown on the right of the diagram above. Consider the closed network $\mathrm{Closure}(T^*T)$.

Closure (T^*T) =

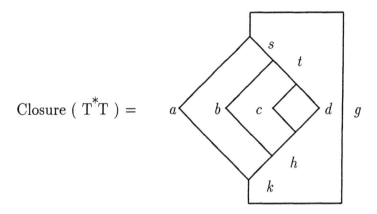

We compute

$$\text{Closure}(T^*T) = \delta_{h,t}\Theta(t,d,c)\delta_{s,k}\Theta(t,b,s)\Theta(a,s,g)/(\Delta_t\Delta_s).$$

Thus

$$\left\{ \begin{array}{ccc} c & d & t \\ g & e & f \end{array} \right\}_q \cdot \left\{ \begin{array}{ccc} b & t & s \\ g & a & e \end{array} \right\}_q$$

$$\cdot\Theta(t,d,c)\Theta(t,b,s)\Theta(a,s,g)/(\Delta_t\Delta_s)$$

$$= \sum_j \left\{ \begin{array}{ccc} b & c & j \\ f & a & e \end{array} \right\}_q \cdot \left\{ \begin{array}{ccc} j & d & s \\ g & a & f \end{array} \right\}_q \cdot \left\{ \begin{array}{ccc} c & d & t \\ s & b & j \end{array} \right\}_q$$

$$\cdot\Theta(t,d,c)\Theta(t,b,s)\Theta(a,s,g)/(\Delta_t\Delta_s).$$

It follows that for any choice of s and t so that (t,d,c), (t,b,s), and (a,s,g) are q-admissible:

$$\left\{ \begin{array}{ccc} c & d & t \\ g & e & f \end{array} \right\}_q \cdot \left\{ \begin{array}{ccc} b & t & s \\ g & a & e \end{array} \right\}_q$$

$$= \sum_j \left\{ \begin{array}{ccc} b & c & j \\ f & a & e \end{array} \right\}_q \cdot \left\{ \begin{array}{ccc} j & d & s \\ g & a & f \end{array} \right\}_q \cdot \left\{ \begin{array}{ccc} c & d & t \\ s & b & j \end{array} \right\}_q \cdot$$

This completes the proof. $\square$

5 The Turaev-Viro Invariant

In this last section we explain how to use the normalized $6j$-symbol (in the case that A is a primitive $4r$th root of unity) to give the definition of the Turaev-Viro invariants of 3-dimensional manifolds. Computations of this invariant can be found in the paper [32] and the book [18]. Specialization to the root of unity case is necessary so that the sum in the definition of the invariant is a finite sum (physicist's renormalization).

While the invariants have not distinguished 3-manifolds that cannot be distinguished in other ways, new applications of these invariants are expected. Furthermore, the framework of a topological quantum field theory — into which the Turaev-Viro invariants fit — is quite general, and is currently being explored in its own right. In particular, there is hope that interesting 4-dimensional generalizations can be found, and that the Donaldson invariants, for example, can be defined as state summations [4]. We mention the following interesting problem: Suppose two manifolds of dimension 3 have the same Turaev-Viro invariants, in what ways are they similar? In other words, what qualitative features do the Turaev-Viro invariants distinguish?

5.1 Definition. Fix an integer $r \geq 3$, let $A = e^{\pi i/(2r)}$, and let $q = A^2$. Let M denote a triangulated 3-dimensional closed manifold. Let t denote the number of vertices, let $\{E_1, \ldots, E_u\}$ denote the set of edges, and let $\{T_1, \ldots, T_w\}$ denote the set of tetrahedra of the triangulation. Let $C = \{0, 1/2, \ldots, (r-2)/2\}$ denote the set of colors associated to the integer r. A *coloring of* M is a mapping $f : \{E_1, \ldots, E_u\} \to C$. An *admissible coloring* is a coloring such that for each triangle with edges E_ℓ, E_m, and E_n the triple $(f(E_\ell), f(E_m), f(E_n))$ is a q-admissible triple of colors.

Suppose that a coloring is admissible, and consider a tetrahedron T with colors a, b, c, j, k, n associated to its edges so that the triples (a, b, n), (n, c, k), (a, j, k), and (b, c, j) are admissible, and these are the labels on the bounding triangles of the tetrahedron. Associate the symbol $T^f = \begin{bmatrix} a & b & n \\ c & k & j \end{bmatrix}_q$ to this tetrahedron.

The value associated to the coloring f of the 3-manifold M is the quantity

$$|M|_f = \Delta^{-t} \prod_{m=1}^{u} \Delta_{f(E_m)} \prod_{p=1}^{w} T_p^f$$

where

$$\Delta = \Delta_j^{-1} \sum_{\{k,\ell:(j,k,l)\text{is } q\text{-admis.}\}} \Delta_k \Delta_l.$$

It is a consequence of the orthogonality identity that the quantity

$$\Delta_j^{-1} \sum_{\{k,\ell:(j,k,l)\text{is } q\text{-admis.}\}} \Delta_k \Delta_l$$

is independent of j (see [32, 18] for a proof).

The *Turaev-Viro* invariant of the 3-manifold M is the state sum

$$|M| = \sum_f |M|_f$$

where the sum ranges over all admissible colorings f of the given triangulation.

5.1.1 Theorem (Turaev-Viro [32]). *The value $|M| \in \mathbf{C}$ is independent of the triangulation chosen; as such it is an invariant of the manifold M.*

Proof. The Pachner Theorem for triangulations of 3-manifolds, states that any two triangulations of a given 3-manifold are related by a sequence of the two moves depicted below where the figure on the left can be manipulated to the figure on the right and *vice versa.*

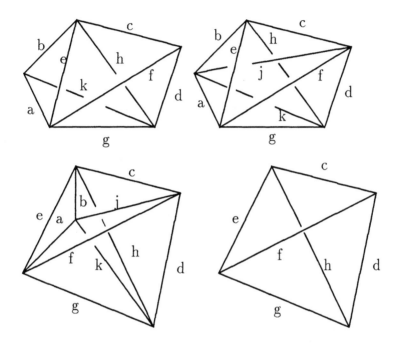

In the first figure two tetrahedra with edges (a, b, e, g, h, k) and (c, d, e, f, g, h) are glued along their common triangular face (g, h, e). Then an edge labelled j is inserted, so the polyhedron is the union of 3 tetrahedra: (a, b, c, e, f, j), (b, c, d, h, j, k), and (a, d, f, g, j, k). The Elliott-Biedenharn identity

$$
\begin{bmatrix} c & d & h \\ g & e & f \end{bmatrix}_q \cdot \begin{bmatrix} b & h & k \\ g & a & e \end{bmatrix}_q
$$

$$
= \sum_j \Delta_j \begin{bmatrix} b & c & j \\ f & a & e \end{bmatrix}_q \cdot \begin{bmatrix} j & d & k \\ g & a & f \end{bmatrix}_q \cdot \begin{bmatrix} c & d & h \\ k & b & j \end{bmatrix}_q
$$

gives that the Turaev-Viro invariant $|M|$ remains unchanged under such a move on triangulations.

In the second figure, the tetrahedron with edges (c, d, e, f, g, h) is subdivided into the union of 4 tetrahedra by adding a vertex with incoming edges (a, b, j, k). In order to prove that $|M|$ remains invariant under such moves on triangulations we prove the following formula holds:

$$\begin{bmatrix} c & d & h \\ g & e & f \end{bmatrix}_q = \Delta^{-1} \sum_{a,b,j,k} \Delta_a \Delta_b \Delta_j \Delta_k \cdot$$

$$\begin{bmatrix} b & c & j \\ f & a & e \end{bmatrix}_q \cdot \begin{bmatrix} j & d & k \\ g & a & f \end{bmatrix}_q \cdot \begin{bmatrix} c & d & h \\ k & b & j \end{bmatrix}_q \cdot \begin{bmatrix} a & b & e \\ h & g & k \end{bmatrix}_q$$

The identity above is called (*). The following pictures, which we explain following their presentation, will help motivate the proof.

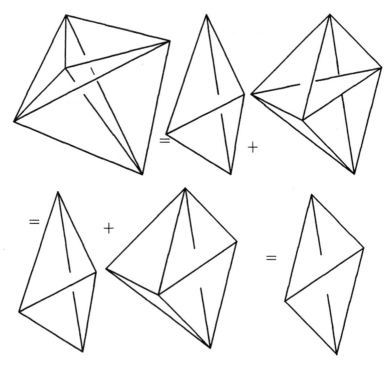

In the top figure, one of the four tetrahedra is considered independently of the other three. Then the union of the three tetrahedra (glued along an edge) is recomposed as the union of two tetrahedra (glued along a face) via the move to triangulations corresponding to the Elliott-Biedenharn identity. The separate tetrahedron is then reglued to one of the tetrahedra in the union of two along three faces. Roughly speaking, we use orthogonality to eliminate the two tetrahedra on the left of the figure. In fact, orthogonality is not quite powerful enough to complete the proof, but orthogonality together with the fact that the element

$$\Delta = \Delta_j^{-1} \sum_{\{k,\ell:(j,k,l)\text{is } q\text{-admis.}\}} \Delta_k \Delta_l.$$

is independent of j will be enough for the proof. Algebraic details follow:

$$\text{RHS}(*) = \Delta^{-1} \sum_a \sum_b \sum_k \Delta_a \Delta_b \Delta_k.$$

$$\begin{bmatrix} c & d & h \\ g & e & f \end{bmatrix}_q \cdot \begin{bmatrix} b & h & k \\ g & a & e \end{bmatrix}_q \cdot \begin{bmatrix} a & b & e \\ h & g & k \end{bmatrix}_q$$

(by the Elliott-Biedenharn identity)

$$= \begin{bmatrix} c & d & h \\ g & e & f \end{bmatrix}_q \Delta^{-1} \sum_a \sum_b \Delta_a \Delta_b \Delta_e^{-1}$$

(by orthogonality)

$$= \text{LHS}(*)$$

The last step follows from $\Delta = \Delta_e^{-1} \sum \Delta_a \Delta_b$ since the sum is taken over all the q-admissible colorings.

It is instructive to reproduce the proof in the geometry of the dual skeleton, in terms of the shadow-world diagrams, or in terms of movies. Any one of these geometrical processes helps encode the calculations.

5.1.2 Definition (Kauffman-Lins [18]). In Kauffman-Lins the following alternative definition of the Turaev-Viro invariant is presented. First a triangulation of a 3-manifold is chosen and dualized. The vertices of the dual correspond to the tetrahedra of the triangulation, the edges in the dual correspond to the faces of the triangulation, and the faces in the dual correspond to edges in the triangulation. Colors are associated to the faces of the dual in such a way that three colors coincident to an edge form a q-admissible triple. To a vertex, at which six faces meet, the value of a tetrahedral spin network is associated. The spins on the edges of the tetrahedral network are the colors associated to the faces of the dual, and the edges of the dual correspond to the vertices of the tetrahedral network. To an edge in the dual with colors a, b and j coincident the value $\Theta(a, b, j)$ is associated. Since these form a q-admissible triple, the value of Θ is defined and is non-zero. To each colored face in the dual we associate the value Δ_j where j is the color on the face.

Thus a *state*, S, of the triangulation is an assignment of colors to the "faces" of the dual such that the colors coincident at an edge form a q-admissible triple. (The term "face" is in quotes above because as the dual complex to the triangulation is deformed, some of the 2-dimensional pieces may not be 2-cells. For example, the Matveev bubble move introduces an annular face.) To such a state, we have $\mathrm{TET}(v|S)$ the value of the tetrahedral spin net associated to each a vertex v dual to the given cell. To an edge, E, with colors (a, b, j) coincident, we associate the value $\Theta(E|S) = \Theta(a, b, j)$. And to a face, f, with color $j(f) = j(f|S)$, associate the value $\Delta_{j(f)}$. Let $\chi(f)$ denote the Euler characteristic of the 2-dimensional face f where the term face is interpreted as above.

If a given edge, e, forms a simple closed curve with no vertices from the dual complex, then let $\chi(e) = 0$; if the edge has a vertex, then let $\chi(e) = 1$. With these conventions, a state sum is defined by the formula

$$|M|_{KL} = \frac{\Delta^{-t+1} \sum_S \left(\Pi_v \mathrm{TET}(v|S) \Pi_f \Delta_{j(f|S)}^{\chi(f)} \right)}{(\Pi_e \Theta(e|S)^{\chi(e)})}.$$

The proof that this does not depend on the choice of triangulations is given in [18]; the proof depends on expressing the tetrahedral networks in terms of the $6j$-symbol. In Kauffman-Lins a nice glimpse of the shadow world is presented, as well.

5.1.3 Theorem (Piunikhin [28]). *The Kauffman-Lins definition and the Turaev-Viro definitions coincide.*

Proof. This is a computation dependent on the definition of the $6j$-symbol $\begin{bmatrix} a & b & c \\ d & e & f \end{bmatrix}_q$. $\square$

5.2 Epilogue.   In the current work we have presented the diagrammatics of the classical and quantum representation theory with topological applications in mind. One major focus has been the Clebsch-Gordan theory and the explicit construction of the $6j$-symbol in all three cases — classical, generic quantum, and quantum root of unity. In the process of developing this theory, we have touched on some other topological aspects that deserve mentioning.

The Jones polynomial [12], which is an invariant of knotted and linked curves in 3-dimensional space, is the starting point of the quantum topology invariants. The bracket identity leads directly to a definition, and this construction can be found in [16], for example.

The quotient of the quantum group $U_q(sl(2))$, when A is a primitive 4rth root of unity, by relations $E^r = F^r = 0$ and $K^{4r} = 1$ has the structure of a modular ribbon Hopf algebra as defined by Reshetikhin-Turaev [29]. Rather than explicitly describing this structure we have worked with the algebra via its representations. However, in doing so, we have verified that the ribbon structure is present. The modular ribbon Hopf algebra gives the Reshetikhin-Turaev invariant, and Lickorish [23] has presented the definition of the invariant in a diagrammatic form. From the present point of view, there is not much more that needs to be done to get to that formulation. The details of that construction can also be found in [16].

The Turaev-Viro invariant is an example of a topological quantum field theory (TQFT), namely a functor from the category of smooth manifold cobordisms to the category of Hilbert spaces. A great deal of effort is currently being exerted towards finding new extended topological field theories, and towards finding higher dimensional analogues. One formulation of the higher dimensional theories is found in Lawrence [22]; from that point of view the structure that is associated to a 3-manifold is a "3-algebra," and the quantum $6j$-symbol gives an explicit construction of such an algebra. In dimension 4, an example of a 4-algebra would give rise to a state sum invariant of the type constructed here. There are very good reasons for searching for such higher dimensional invariants. Evidence for their existence is given by the solutions to the tetrahedral equation [35] which is an analogue of the Yang-Baxter equation. Furthermore, non-trivial examples of these higher algebraic structures will give new and in-depth meaning to the well-known algebraic structures. We have seen this happen already in the passage from classical $sl(2)$ to quantum $sl(2)$. For computations of the Turaev-Viro invariants see [18].

Finally, there are deep connections to theoretical physics that require much further study from the mathematical, theoretical, and experimental sides. The mathematical aspect that is the most problematic is the definition of functional integration — which might be thought of as a continuous version of the state sum method. A rigorous definition of the functional integral will lead to analytic interpretations of the algebra and of the topology, and such intepretations will certainly shed light on all of the aspects of the theory.

References

[1] Alexander, J.W., *The Combinatorial Theory of Complexes*, Annals of Math. 31 (1930), 175-186.

[2] Biedenharn, L. C. and Louck, J. D., "The Racah-Wigner Algebra in Quantum Theory," Encyclopedia of Mathematics, Addison-Wesley (1981).

[3] Brink, D. M. and Satchler, G. R., "Angular Momentum," Oxford University Press 2nd edition, (1975).

[4] Crane, L. and Frenkel, Igor, *Four Dimensional Topological Quantum Field Theory, Hopf Categories, and Canonical Basis*, J. Math. Phys. 35 (10), (Oct 1994), p. 5136.

[5] Drinfel'd, V. G., *Quantum Groups*, Proc. ICM-86 (Berkeley), vol.1, Amer.Math.Soc., (1987), 798-820.

[6] Drinfe'ld, V. G., *Quasi-Hopf Algebras and Knizhnik-Zamolodchikov Equations*, Research Reports in Physics, Problems of Modern Quantum Field Theory, (Circa 1990).

[7] Goodman, F. M. and Wentzl, H., *The Temperley-Lieb Algebra at Roots of Unity*, Pacific Journal of Math. Vol 161, No.2 (1993), 307-334.

[8] Humphreys, J. E., "Introduction to Lie Algebras and Representation Theory," Springer-Verlag (New York 1972).

[9] Jacobson, N., "Basic Algebra II," W. H. Freeman Co. (San Francisco 1980).

[10] Jimbo, M., *A q-Difference Analogue of $U(g)$ and the Yang-Baxter Equation*, Letters Math. Phys. 10 (1985), 63-69.

Reprinted in Jimbo, M., "Yang-Baxter Equation in Integrable Systems," World Scientific Publishing Co., (Singapore 1989).

[11] Jimbo, M., "Yang-Baxter Equation in Integrable Systems," World Scientific Publishing Co., (Singapore 1989).

[12] Jones, V. F. R., *Hecke Algebra Representations of Braid Groups and Link Polynomials*, Ann. of Math. 126 (1987), 335-388. Reprinted in Kohno "New Developments in the Theory of Knots," World Scientific Publishing (Singapore 1989).

[13] Jones, V. F. R., *Index for Subfactors,* Inventiones Math. 72 (1983), 1-25. Reprinted in Kohno "New Developments in the Theory of Knots," World Scientific Publishing (Singapore 1989).

[14] Kauffman, L., *Spin Networks and the Jones Polynomial,* Twistor Newsletter, No. 29 (8 November 1989), Mathematics Institute, Oxford, 25-30.

[15] Kauffman, L., $SL(2)_q$-*Spin Networks,* Twistor Newsletter, No. 32 (12 March 1991), Mathematics Institute, Oxford, 10-14.

[16] Kauffman, L., "Knots and Physics," World Scientific Publishing (Singapore 1991).

[17] Kauffman, L., *Map Coloring, q-Deformed Spin Networks, and the Tureav-Viro Invariants for 3-Manifolds*, International Journal of Modern Physics B, Vol. 6, Nos. 11 & 12, (1992) 1765-1794.

[18] Kauffman, L. and Lins, S., "The Temperley-Lieb Algebra Recoupling Theory and Invariants of 3-Manifolds," Annals of

Math Studies Vol. 134, Princeton University Press (Princeton 1994).

[19] Kirillov, A. N. and Reshetikhin, N. Yu, *Representations of the Algebra $U_q(sl(2))$, q-Orthogonal Polynomials and Invariants of Links.* Reprinted in Kohno "New Developments in the Theory of Knots," World Scientific Publishing (1989).

[20] Keller, G., *Fusion Rules for $U_q(sl(2, \mathbf{C}))$, $q^m = 1$,* Letters Math. Phys. 21 (1991), 273-286.

[21] Ko, Ki Hyoung, and Smolinsky, Lawrence, *A Combinatoric Matrix in 3-Manifold Theory,* Pacific Journal of Math., Vol 149, No. 2, (1991), 319-336.

[22] Lawrence, R. L., *Algebras and Triangle Relations,* in "Topological and Geometric Methods in Field Theory" eds J. Michelsson, O. Pekonen, World Scientific Publishing (1992), 429-447; *Algebras and Triangle Relations,* to appear in J. Pure Appl. Alg. 100 (1995).

[23] Lickorish, W. B. R., *The Skein Method for Three-Manifold Invariants,* Journal of Knot Theory and Its Ramifications, Vol 2, No. 2 (1993), 171-194.

[24] Masbaum, G. and Vogel, P., *3-valent graphs and the Kauffman bracket,* Pacific Journal of Math. (1994), 361-381.

[25] Matveev, S.V., *Transformations of Special Spines and the Zeeman Conjecture,* Math USSR Izvestia, Vol. 31 (1988), 423-434.

[26] Pachner, U., *PL homeomorphic manifolds are equivalent by elementary shelling,* Europ. J. Combinatorics Vol. 12 (1991), 129-145.

[27] Penrose, R., *Applications of Negative Dimensional Tensors*, in Welsh, "Combinatorial Mathematics and its Applications," Academic Press (1971).

[28] Piunikhin, S., *Turaev-Viro and Kauffman-Lins Invariants for 3-Manifolds Coincide*, Journal of Knot Theory and Its Ramifications, Vol 1, No. 2 (1992), 105-135.

[29] Reshetikhin, N. and Turaev, V., *Invariants of 3-Manifolds Via Link Polynomials*, Inventiones Math. 103 (1991), 547-597.

[30] Shnider, S. and Sternberg, S., "Quantum Groups: from coalgebras to Drinfeld algebras a guided tour," International Press Publications (Cambridge, MA 1993).

[31] Sternberg, S., "Group Theory and Physics," Cambridge University Press (Cambridge, UK 1994).

[32] Turaev, V. and Viro, O., *State Sum Invariants of 3-Manifolds and Quantum 6J-Symbols*, Topology Vol. 31, No 4 (1992), 865-902.

[33] Weyl, Hermann, "The Classical Groups," Princeton University Press (Princeton 1946).

[34] Witten, E., *Quantum Field Theory and the Jones Polynomial*, Comm. Math. Phys. 121 (1989), 715-750. Reprinted in Kohno, "New Developments in the Theory of Knots," World Scientific Publishing (Singapore 1989).

[35] Zamolodchikov, A. B., *Tetrahedron equations and the relativistic S-matrix of straight-strings in 2 + 1-dimensions*, Comm. Math. Phys. 79 (1981), 489-505; Reprinted in Jimbo,

"Yang-Baxter Equation in Integrable Systems," World Scientific Publishing Co., (Singapore 1989).